ILLUSTRATED HANDBOOK OF ELECTRONIC TABLES, SYMBOLS, MEASUREMENTS AND VALUES

ILLUSTRATED HANDBOOK OF ELECTRONIC TABLES, SYMBOLS, MEASUREMENTS AND VALUES

Raymond H. Ludwig

PARKER PUBLISHING COMPANY, Inc.
WEST NYACK, NEW YORK

Library of Congress Cataloging in Publication Data
Ludwig, Raymond H
    Illustrated handbook of electronic tables, symbols,
measurements, and values.

    Includes index.
    1. Electronics--Handbooks, manuals, etc. I. Title.
TK7825.L83   621.381   77-819
ISBN 0-13-450973-0

Printed in the United States of America

To Dawn and Dad

About This Instant Reference Data . . . .

Almost every profession has its own language, symbols, and specialized terminology. In Electronics, this is especially true. We must know and apply a countless number of facts, techniques, formulas, tables, rules, component identification characteristics, wiring codes, abbreviations, symbols, and meanings, blending all of this knowledge with the latest applications and techniques.

This "Illustrated Handbook" provides the electronics profession with an invaluable source of usable, practical information. No longer will you have to waste valuable time searching through a variety of sources for the essential information you need. This book will be as helpful to those in the field of Electronics as is the dictionary to most other people.

Each section provides a broad range of practical electronic reference data arranged into seven memory-rejuvenating chapters, each concentrating on standardized electronic data, including symbols, abbreviations, and related reference data. In order to make your research time shorter yet, the contents within each chapter are sub-divided into specific electronic data sections, each of which is identified by its chapter number and special number grouping. The valuable data found within each sub-section has, wherever possible, been alphabetized.

The first chapter provides a gold mine of facts you need to know concerning the language used in electronics. Information on schematic symbols, logic symbols, abbreviations, mathematical signs and symbols, or transistor and vacuum tube letter designations can be located and put to use with a minimum investment of your time. For example, if you are concerned with terms and nomenclature related to schematic symbols for a particular transistor, then first turn to the solid state schematic symbols listed in Chapter 1 for the symbols employed (Section 1.6), then to the section or sections of Chapter 1 covering transistor letter symbols (Sections 1.7-0 to 1.7-8). In doing this, you will not only become familiar with other schematic symbols which may identify the same transistor, but also with the knowledge of whether the letters are describing the DC, RMS, instantaneous, maximum or peak values relating to voltage, current, resistance, amplification, conductance, power or capacitance.

The ready reference, quick review format of Chapter 2 provides all the essential facts and rules for electronic calculations. It includes arithmetic, algebra, geometry, trigonometry, complex numbers and vectors, logarithms, and advanced mathematics. The related facts and rules pertinent to each of the seven main headings are easily identified and readily found since each has its own number grouping. If, for example, you need the mathematic process for multiplying vectors in rectangular form, you would turn to Chapter 2 dealing with mathematics for electronics, Section 5 concerning rectangular vector form, and sub-section 4B illustrating the multiplication process (2.5-4B).

Chapter 3 has pulled together all the important rules, laws, characteristics, facts, and formulas concerning 23 separate electronic categories. These 23 main headings range alphabetically from Admittance to Wavelength and Waveguides. Each numbered main heading serves for grouping the numerous related facts, characteristics, etc. applicable to that subject. An additional bonus in this chapter is that each section and sub-section points out which facts, formula, rule, etc. apply to AC applications, DC applications, or AC and DC applications. For instance, all the data for AC applications are found in Section 3.2 or alphabetically under Alternating Current, while DC applications are in Section 3.8 or in alphabetical arrangement under Direct Current. In those areas where AC and DC application is possible it is so stated, thereby avoiding any misconceptions.

The key to success in the sophisticated field of electronics is often a blend of knowledge and common sense. Chapter 4 will supply you with both and will eliminate possible false assumptions easily made due to non-standardizations which result from rapid technology advances. The valuable information found within this chapter ''tells it how it is'' concerning all types of capacitors, circuit breakers and fuses, transformers, axial lead resistors, color codings and other identifying features, semiconductor device packaging, simple testing methods, plus vacuum tube data.

Chapter 5 covers a broad range of time-saving information normally given in table form. Portions of this material may be found separately in a variety of books and pamphlets, but now all this material has been pulled together and organized for you in one place. Some of this hard-to-get-together data includes: AWG wire specifications, decibels, drill, tap, and machine screw data, pilot lamp specifications, RF cable specifics along with TV channel allocations.

Chapter 6 provides hundreds of alphabetized conversion factors that are absolutely essential throughout the highly specialized field of electronics. You will find them invaluable when working with acceleration, angles, area, electrical units, energy, force, heat, length, magnetic units, mass, power, pressure, temperature, velocity, volume, or weight.

Chapter 7 is the electronic analysis portion of this book. A common 3-tube transformerless amplifier is dissected and reconstructed stage by stage, component by component, using design rules, formulas, facts, and technical reasons necessary to pick the values for components used. In doing this you not only will see the hows and whys of circuitry design and operation, but will also be able to relate component names, purposes, and design characteristics to a limitless number of other circuits you see daily.

To sum up, the practical value offered by this book is almost limitless. It provides in one volume, an extraordinary range of important reference data needed by the electronic professional.

Raymond H. Ludwig

CONTENTS

3. KEY FACTS AND FORMULAS FOR ELECTRONIC PROBLEM SOLVING
(Cont.)

CHAPTER 1

Directory of Electronic Symbols

Communication between individuals is actually a two-way street because the speaker and listener must talk the same language to avoid any misconceptions. Sometimes what you are trying to convey is known by someone else as being something different and may lead to a communications gap.

The terms, schematic symbols, subscripts, and abbreviations commonly employed in the vast field of electronics make up a specialized language, a language full of possible communication gaps. One reason for this is because the standards governing our language often have become modified within an industry, so that each company has its own individual language.

The following ten sections of Chapter 1 will close any electronic communication gaps that may exist in your language, by quickly and clearly relating standardized data to those meaning the same thing.

To aid you further, all of the illustrations and data have been alphabetized within each particular section. For example, Section 1.1 on circuit element schematic symbols has three groupings for capacitor symbols; one being *fixed*, another being for *electrolytic*, and the third group for variable types. Illustrated within each group are the symbols commonly used in schematic drawings.

1.1   CIRCUIT ELEMENT SCHEMATIC SYMBOLS

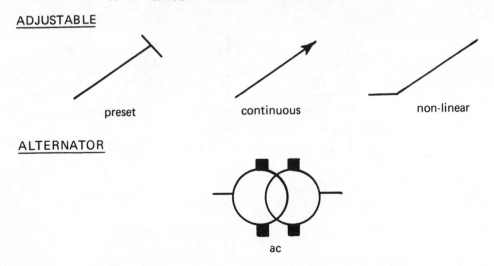

ADJUSTABLE

preset   continuous   non-linear

ALTERNATOR

ac

AMPERAGE

DC   AC

AMPLIFIER

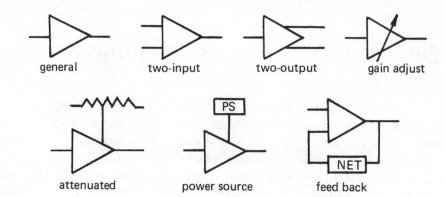

general   two-input   two-output   gain adjust

attenuated   power source   feed back

ANTENNA

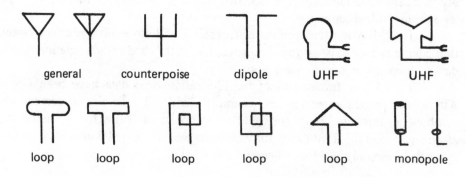

general   counterpoise   dipole   UHF   UHF

loop   loop   loop   loop   loop   monopole

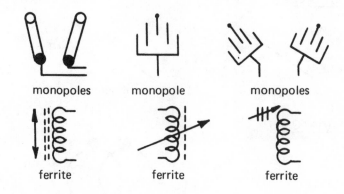

monopoles   monopole   monopoles

ferrite   ferrite   ferrite

ARRESTER-LIGHTNING

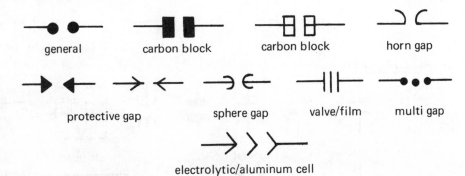

general   carbon block   carbon block   horn gap

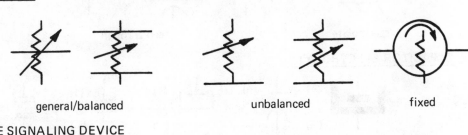

protective gap   sphere gap   valve/film   multi gap

electrolytic/aluminum cell

ATTENUATOR

general/balanced   unbalanced   fixed

AUDIBLE SIGNALING DEVICE

bell   buzzer   bell & buzzer   speaker   telegraph sounder
                                                      siren
                                                      horn

BRAKE, CLUTCH

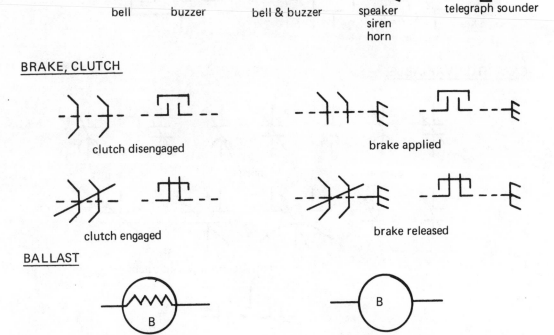

clutch disengaged   brake applied

clutch engaged   brake released

BALLAST

B   B

BATTERY

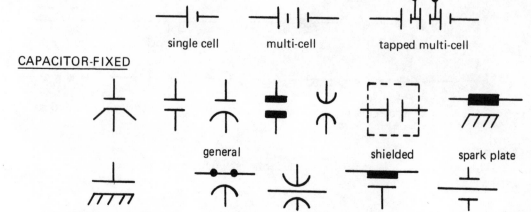

single cell   multi-cell   tapped multi-cell

CAPACITOR-FIXED

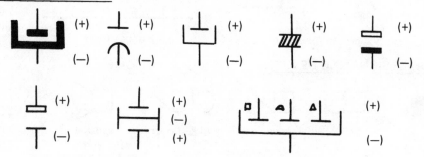

general   shielded   spark plate

spark plate   feed through

CAPACITOR-ELECTROLYTIC

CAPACITOR-VARIABLE

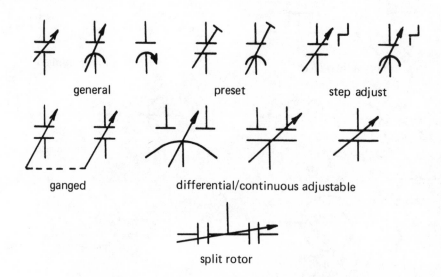

general   preset   step adjust

ganged   differential/continuous adjustable

split rotor

CELLS-PHOTOSENSITIVE

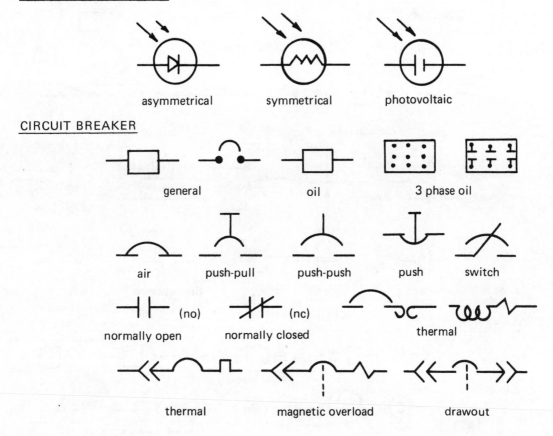

asymmetrical   symmetrical   photovoltaic

CIRCUIT BREAKER

general   oil   3 phase oil

air   push-pull   push-push   push   switch

(no)   (nc)   thermal

normally open   normally closed

thermal   magnetic overload   drawout

COAXIAL CABLE

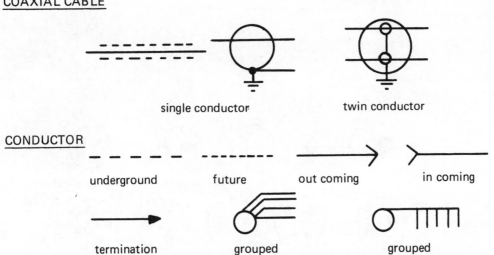

single conductor   twin conductor

CONDUCTOR

underground   future   out coming   in coming

termination   grouped   grouped

CONTACTS

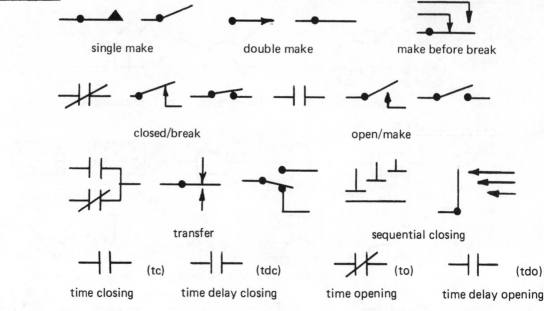

single make   double make   make before break

closed/break   open/make

transfer   sequential closing

time closing   (tc)   time delay closing   (tdc)   time opening   (to)   time delay opening   (tdo)

CONNECTOR

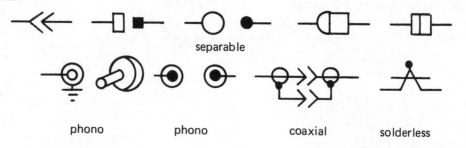

separable

phono   phono   coaxial   solderless

CONNECTOR-MALE PLUG

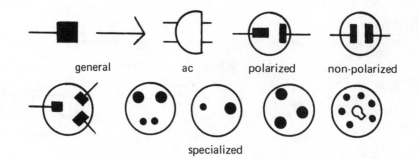

general   ac   polarized   non-polarized

specialized

CONNECTOR-FEMALE

general ac polarized non-polarized

specialized

COUPLERS-DIRECTIONAL

general
aperture

"E" plane
30 db loss

(E)
30 db

loop coupling
30 db loss

30 db

probe coupling
30 db loss

30 db

resistive coupling
30 db loss

30 db

COUPLING-HIGH FREQUENCY

loop to space loop to guide probe to guide probe to guide

loop from coaxial to
circular waveguide

probe from coaxial to
rectangular waveguide

CRYSTAL-DETECTOR

general

CRYSTAL

piezoelectric   general

FUSE

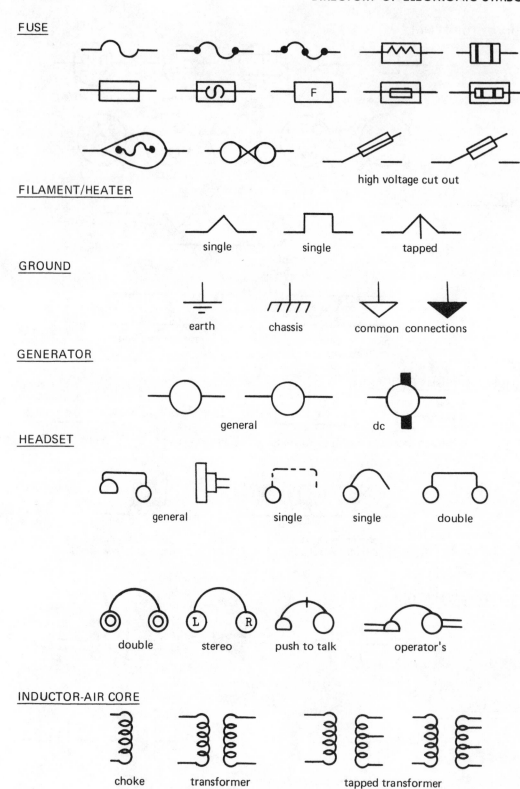

high voltage cut out

FILAMENT/HEATER

single single tapped

GROUND

earth chassis common connections

GENERATOR

general dc

HEADSET

general single single double

double stereo push to talk operator's

INDUCTOR-AIR CORE

choke transformer tapped transformer

INDUCTOR-MAGNETIC/IRON CORE

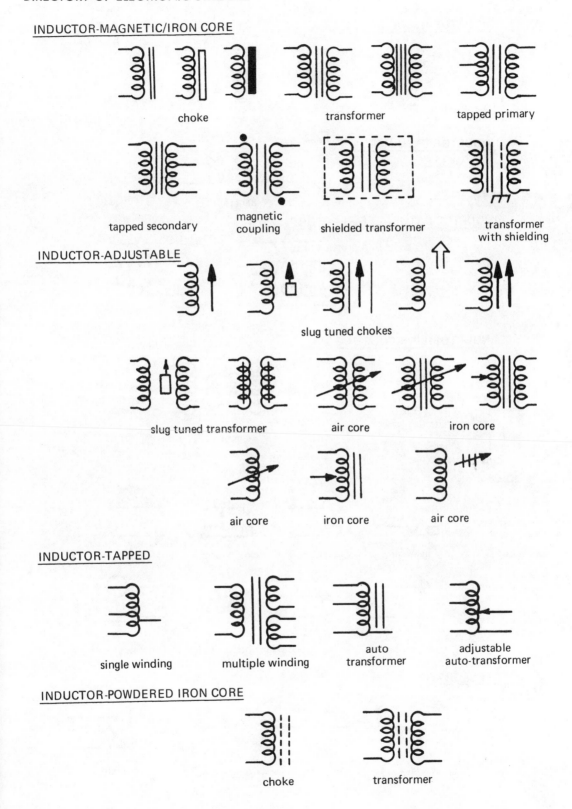

choke   transformer   tapped primary

tapped secondary   magnetic coupling   shielded transformer   transformer with shielding

INDUCTOR-ADJUSTABLE

slug tuned chokes

slug tuned transformer   air core   iron core

air core   iron core   air core

INDUCTOR-TAPPED

single winding   multiple winding   auto transformer   adjustable auto-transformer

INDUCTOR-POWDERED IRON CORE

choke   transformer

INDUCTOR-RESOLVER

general

INDUCTOR-DYNAMOTOR

general

INDUCTOR-SATURABLE REACTOR

general

INDUCTOR-LINK COUPLING

general

JACK

general   open circuit   closed circuit   phono

microphone   coaxial/phono   coaxial

KEY-TELEGRAPH

simple   with shorting   open circuit
                switch   pole changing

LAMP-BULBS

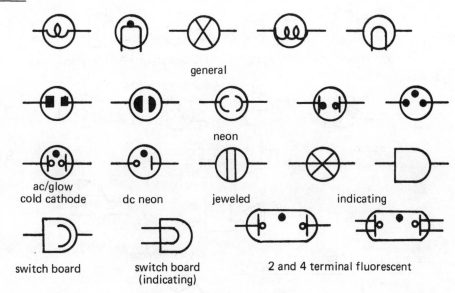

general

neon

ac/glow
cold cathode   dc neon   jeweled   indicating

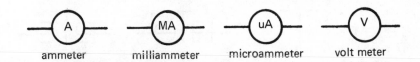

switch board   switch board   2 and 4 terminal fluorescent
(indicating)

METER

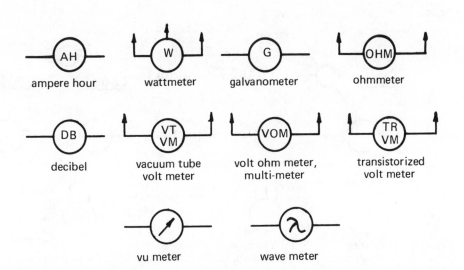

ammeter   milliammeter   microammeter   volt meter

ampere hour   wattmeter   galvanometer   ohmmeter

decibel   vacuum tube   volt ohm meter,   transistorized
volt meter   multi-meter   volt meter

vu meter   wave meter

MICROPHONE

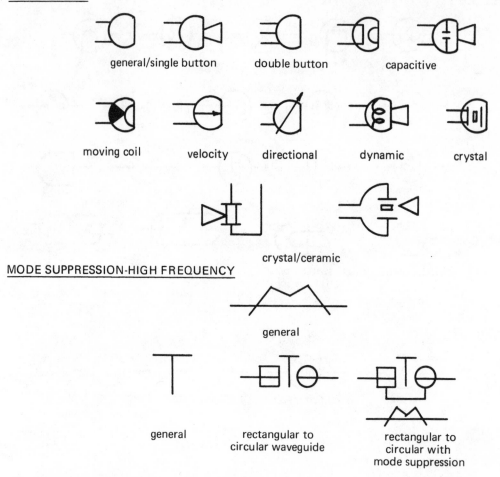

general/single button   double button   capacitive

moving coil   velocity   directional   dynamic   crystal

crystal/ceramic

MODE SUPPRESSION-HIGH FREQUENCY

general

general   rectangular to circular waveguide   rectangular to circular with mode suppression

MOTOR

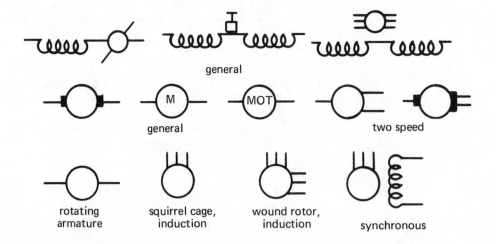

general

general   two speed

rotating armature   squirrel cage, induction   wound rotor, induction   synchronous

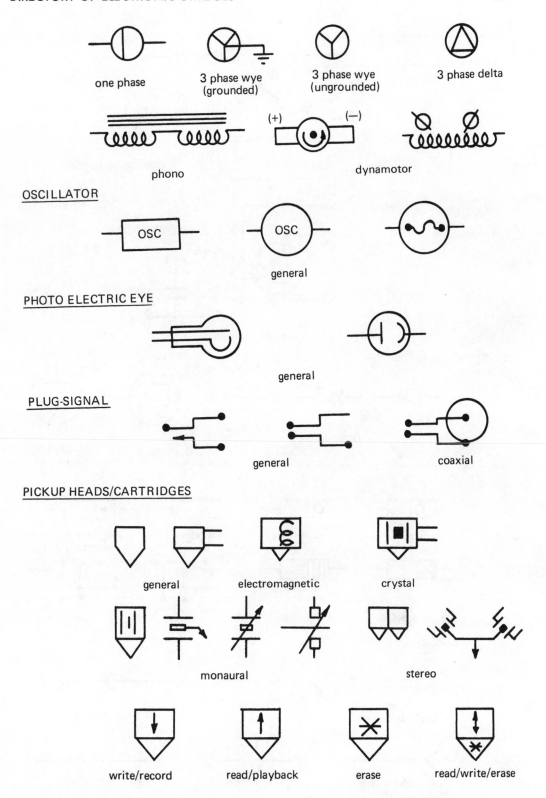

one phase

3 phase wye (grounded)

3 phase wye (ungrounded)

3 phase delta

(+) (−)

phono

dynamotor

OSCILLATOR

OSC

OSC

general

PHOTO ELECTRIC EYE

general

PLUG-SIGNAL

general

coaxial

PICKUP HEADS/CARTRIDGES

general

electromagnetic

crystal

monaural

stereo

write/record

read/playback

erase

read/write/erase

RECTIFIER-FULL WAVE

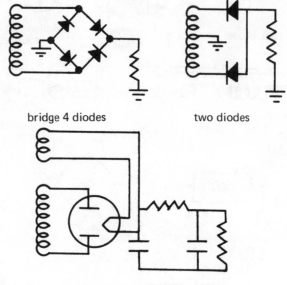

bridge 4 diodes   two diodes

vacuum tube with filter

RECTIFIER-HALF WAVE

general
(solid state)

indirectly heated   directly
heated

RELAY

ac ringing   fast operate   fast release   magnetically
polarized

slow operate   slow release   make   break

RELAY-additional facts

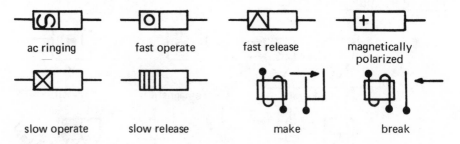

over
(voltage/current)

under
(voltage/current)

differential current

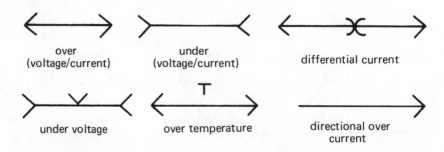

under voltage   over temperature   directional over
current

RESISTOR

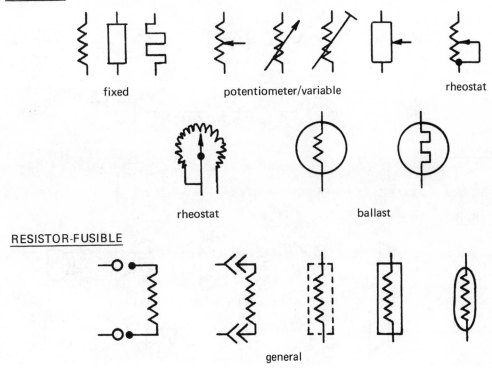

fixed   potentiometer/variable   rheostat

rheostat   ballast

RESISTOR-FUSIBLE

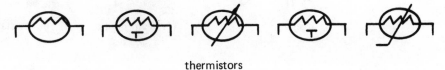

general

RESISTOR-TEMPERATURE COMPENSATED

thermistors

RESISTOR-VOLTAGE DEPENDENT

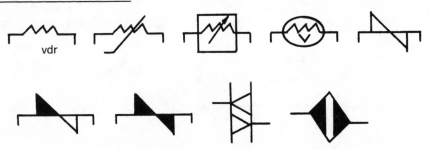

vdr

RESISTOR-LIGHT DEPENDENT

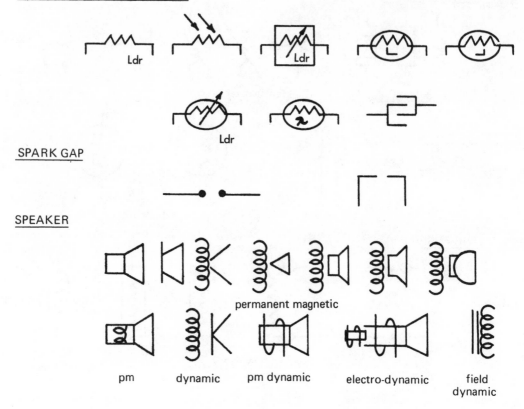

SPARK GAP

SPEAKER

permanent magnetic

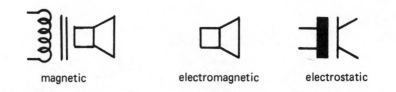

pm   dynamic   pm dynamic   electro-dynamic   field dynamic

magnetic   electromagnetic   electrostatic

SWITCH-SINGLE POLE

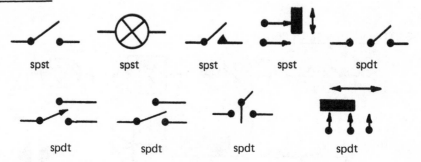

spst   spst   spst   spst   spdt

spdt   spdt   spdt   spdt

SWITCH-DOUBLE POLE

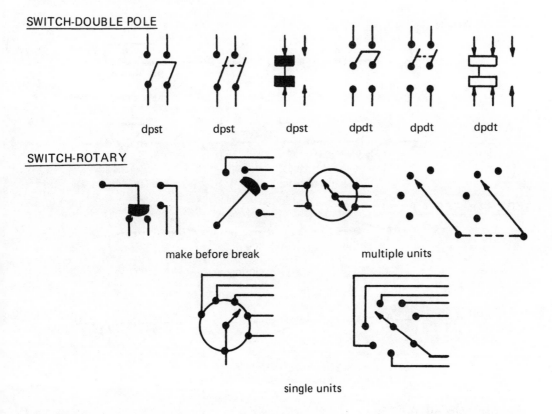

dpst   dpst   dpst   dpdt   dpdt   dpdt

SWITCH-ROTARY

make before break   multiple units

single units

SWITCH-PUSH/PULL

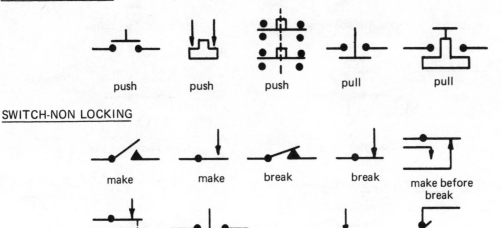

push   push   push   pull   pull

SWITCH-NON LOCKING

make   make   break   break   make before break

two circuit   transfer

SWITCH-LOCKING

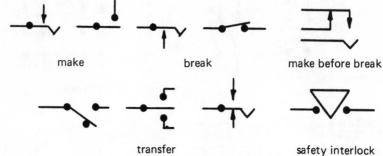

make   break   make before break

transfer   safety interlock

SWITCH-TIME DELAY

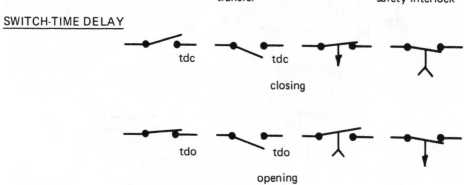

tdc   tdc

closing

tdo   tdo

opening

SWITCH-FLOW ACTIVATED

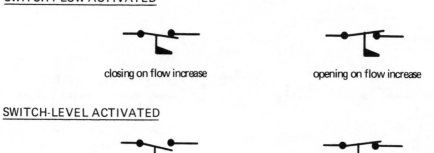

closing on flow increase   opening on flow increase

SWITCH-LEVEL ACTIVATED

closing with rising level   opening with rising level

SWITCH-PRESSURE ACTIVATED

closing with rising pressure   closing with rising pressure

SWITCH-TEMPERATURE ACTIVATED

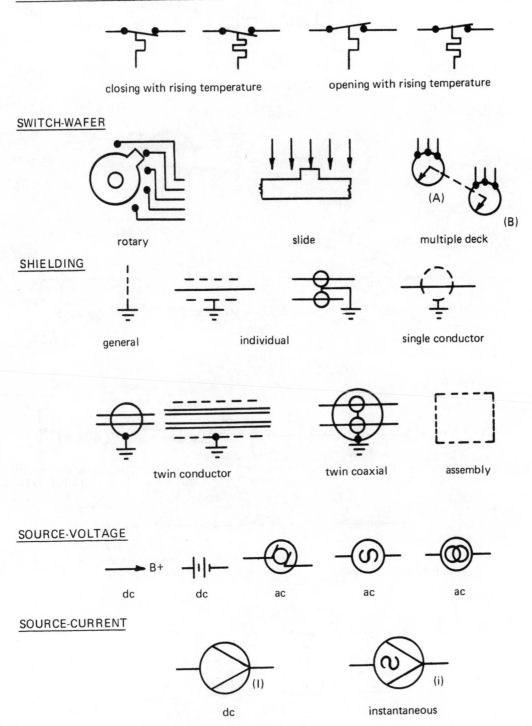

closing with rising temperature   opening with rising temperature

SWITCH-WAFER

rotary   slide   multiple deck

SHIELDING

general   individual   single conductor

twin conductor   twin coaxial   assembly

SOURCE-VOLTAGE

dc   dc   ac   ac   ac

SOURCE-CURRENT

dc   instantaneous

SQUIB

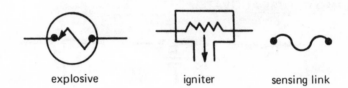

explosive   igniter   sensing link

SYNCHROS

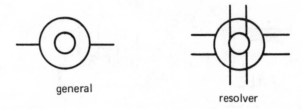

general

resolver

TERMINATION-LINE

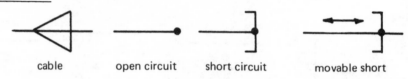

cable   open circuit   short circuit   movable short

TERMINATION-LINE continued

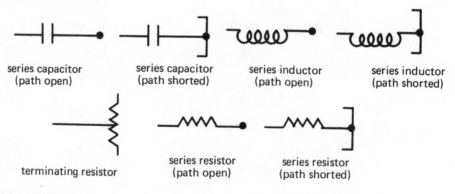

series capacitor series capacitor series inductor series inductor
(path open)   (path shorted)   (path open)   (path shorted)

terminating resistor series resistor series resistor
         (path open)   (path shorted)

TERMINAL BOARD/STRIP

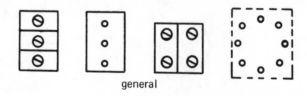

general

THERMOCOUPLE

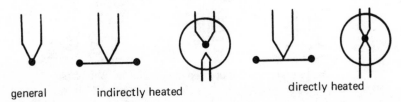

general   indirectly heated   directly heated

THERMO CUTOUT

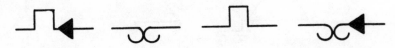

THERMO RELAY

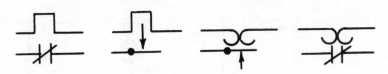

THERMOSTAT CONTACTS

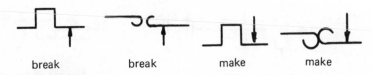

break   break   make   make

THERMOSTAT-NON ADJUSTABLE

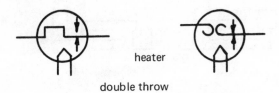

heater

double throw

THERMOSTAT-ADJUSTABLE

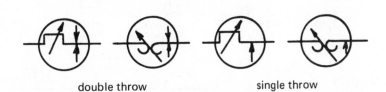

double throw   single throw

TRANSFORMER

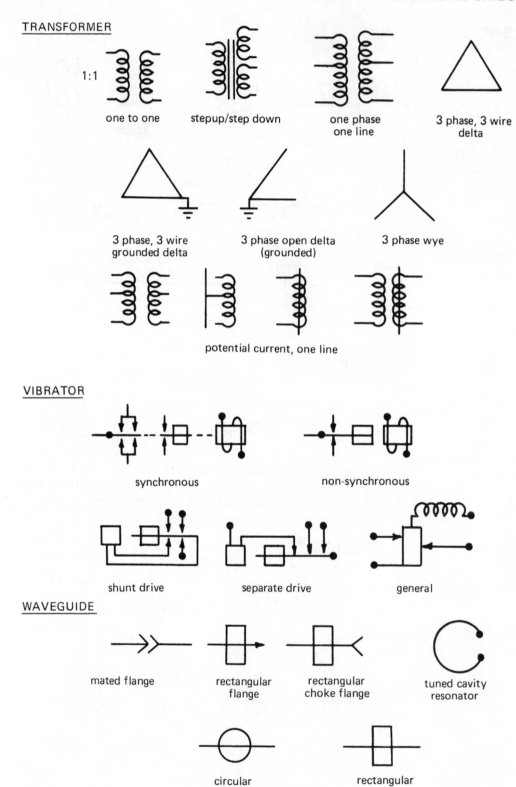

1:1

one to one   stepup/step down   one phase   3 phase, 3 wire
                                   one line   delta

3 phase, 3 wire   3 phase open delta   3 phase wye
grounded delta   (grounded)

potential current, one line

VIBRATOR

synchronous   non-synchronous

shunt drive   separate drive   general

WAVEGUIDE

mated flange   rectangular   rectangular   tuned cavity
                   flange   choke flange   resonator

circular   rectangular

WAVEGUIDE-ROTARY JOINT

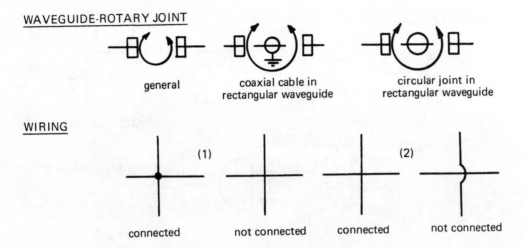

general   coaxial cable in   circular joint in
                 rectangular waveguide   rectangular waveguide

WIRING

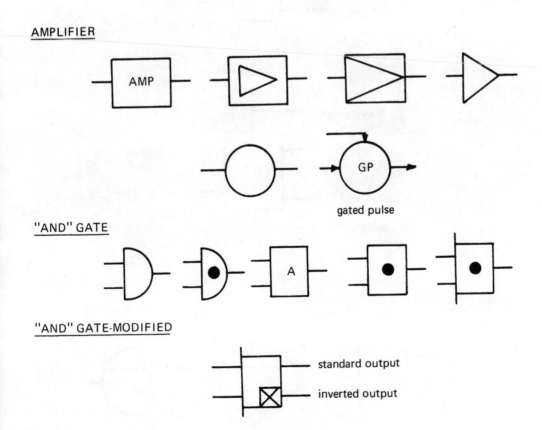

connected   not connected   connected   not connected

(1) connected and not connected (2) connected and not connected
*Schematic diagrams may use either #1 or #2, but not both.

1.2 DIGITAL CIRCUIT LOGIC SYMBOLS

AMPLIFIER

gated pulse

"AND" GATE

"AND" GATE-MODIFIED

standard output

inverted output

"AND NOT"/INHIBITED GATE

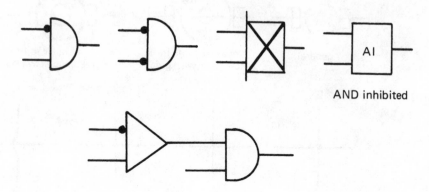

AND inhibited

COMPLEMENTARY FLIP FLOP

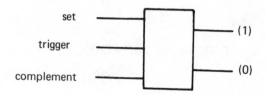

COMPLEMENT FLIP FLOP

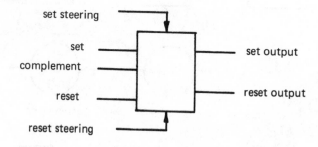

"EXCLUSIVE OR" GATE

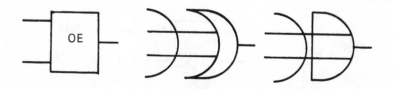

FULL ADDER

HALF ADDER

INVERTER

LATCH FLIP FLOP

"NAND" GATE

"NOR" GATE

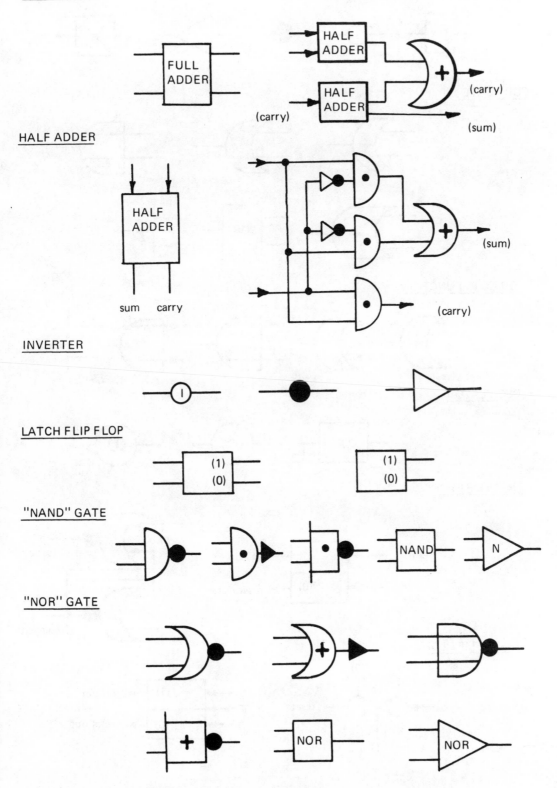

"NOT" FUNCTION

"OR" GATE

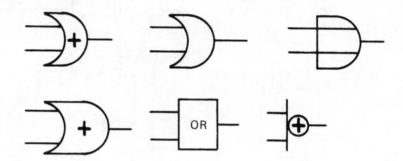

"OR NOT" FUNCTION

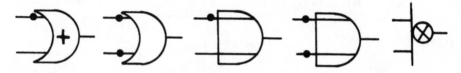

OSCILLATOR

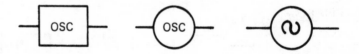

"RS" FLIP FLOP

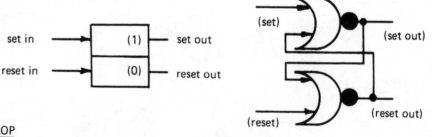

set in ──▶ (1) ── set out
reset in ──▶ (0) ── reset out

(set)   (set out)

(reset)   (reset out)

"RST" FLIP FLOP

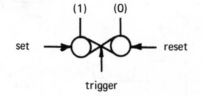

(1)   (0)

set ──▶   ◀── reset

trigger

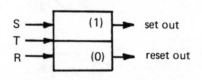

S ──▶ (1) ──▶ set out
T ──▶
R ──▶ (0) ──▶ reset out

"T" FLIP FLOP

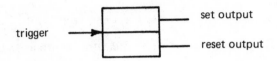

TIME DELAY

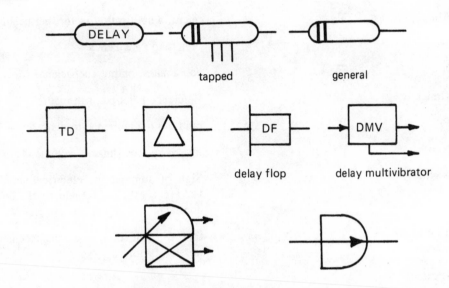

1.3   GREEK ALPHABET WITH SPECIALIZED MEANINGS

NAME	UPPERCASE	LOWERCASE	COMMON USE
Alpha	A	$\propto$, α	angles; absorption area, attenuation constant, coefficients, amplification factor, current gain
Beta	B	β, δ	angles, flux density, coefficients, phase constant, current gain
Gamma	Γ	γ	conductivity, specific gravity, angles, propagation constant
Delta	Δ	δ, ∂	variation, density, angles, increment
Epsilon	E	ϵ	base for natural logarithms, dielectric constant, electrical intensity
Zeta	Z	ζ	impedance, coefficient, coordinates
Eta	H	η	hysteresis coefficient, efficiency, surface charge density, coordinates, intrinsic impedance

Theta	Θ	$\theta, \Theta, \vartheta$	temperature; phase angle, time constant, reluctance
Iota	I	ι	current, vector unit
Kappa	K	κ	dielectric constant, susceptibility, coupling coefficient
Lamba	Λ	λ	wave length, attenuation constant
Mu	M	μ	micro, amplification factor, permeability
Nu	N	ν	reluctivity, frequency
Xi	Ξ	ξ	coordinates, output coefficient
Omicron	O	o	Reference point/point in math
Pi	Π	π	ratio of circumference to diameter = 3.1416
Rho	P	ρ	resistivity, coordinates, volume charge density
Sigma	Σ	σ, ς	Sign of summation, electrical conductivity, leakage coefficient, conductivity, propagation constant
Tau	T	τ	time constant, time phase displacement, transmission factor
Upsilon	Υ	υ	
Phi	Φ	ϕ, φ	magnetic flux, phase angles
Chi	X	χ	Reactance, angles, electrical susceptibility
Psi	Ψ	ψ	dielectric flux, phase difference or angle, coordinates
Omega	Ω	ω	(capital-Ohms) (lowercase-angular velocity)

1.4   LETTER SYMBOL AND ABBREVIATION GLOSSARY

MEANING	ABBREVIATION	SYMBOL
Adjustable	adj, ADJ	
Admittance	————	y, Y
Alternating current	ac, AC	AC
Alternating current—direct current	ac/dc AC/DC	
Ambient	amb	
American Society of Mechanical Engineers	ASME	
American Standards Association	EIA (now USASI)	
American Wire Gage	AWG	
Ampere	a, A, amp, AMP	A

MEANING	ABBREVIATION	SYMBOL
Ampere hour	AMP HR	Ah
Ampere turn	AT	A
Amplification factor	Mu	μ
Amplitude modulation	am, AM	AM
Antenna	ant, ANT	
Antilogarithm	antilog	
Atmosphere	atm	
ATTO (1×10^{-18})	a	
Audio frequency	af, AF	
Automatic frequency control	afc, AFC	
Automatic gain control	agc, AGC	
Automatic volume control	avc, AVC	
Average	avg	
Bandpass frequency of resonant circuits	Af	
Beat frequency oscillator	bfo, BFO	
Bel	B	
Binary coded decimal	BCD	
Bits per second	B/S	
British thermal unit	BTU	Btu
Broadcast	bc, BC	
Calibrate	Cal	
Calorie	CAL	cal
Capacitance	———	c
Capacitive reactance	———	Xc
Capacitor	cap	c
Cathode ray oscilloscope	cro, CRO	
Cathode ray tube	crt, CRT	
Centi (1×10^{-2})	c	
Centigram	cg	
Centimeter (metre is preferred spelling)	———	cm
Centimetre-gram-second	CGS	cgs
Characteristic output impedance	———	Z_0
Charge	———	Q
Circuit	ckt	
Circular mil	c mil	
Clockwise	cw, CW	
Collector	C	
Conductance (old unit was mhos—new unit is siemens)	———	G (mhos) S (siemens)
Conductivity	———	σ
Constant	———	K
Continuous wave	cw	
Counter clockwise	CCW, ccw	
Cosecant	csc	
Cosine	cos	
Cotangent	cot	

MEANING	ABBREVIATION	SYMBOL
Coulomb		C (not Q)
Counter-electromotive force	cemf, CEMF	
Coupling (coefficient of)	———	k
Coupling (resistor-capacitor)	———	R-C
Cubic Centimetre (preferred spelling)	CC	cm^3
Cubic foot	CU FT	ft^3
Cubic foot per minute	CFM	ft^3/min
Cubic foot per second	CFS	ft^3/s
Cubic inch	CU IN	in^3
Cubic Metre (preferred spelling)	CU M	m^3
Cubic millimetre	CU MM	mm^3
Current	———	I
Current (effective-rms value)	I eff	I
Current (instantaneous value)	———	i
Current (maximum value)	———	I max
Cycle (replaced with hertz)	C	c
Cycle per second	(old) cps	Hz
Deci (1x10^{-1})	d	
Decibel	DB	dB
Decibel against 1 mw standard	dBm	
Decibel against 1 watt standard	dBW	
Decibel against 1 volt standard	dBV	
Degrees (Celsius)	———	°C
Degrees (Fahrenheit)	———	°F
Degrees (Kelvins preferred, not degrees Kelvin)	——— ———	K
Deka (1x10^{+1})	da	
Density, flux in gauss	b	β
Diameter	d, dia, DIA	
Digital voltmeter	DVM	
Diode transistor logic	DTL	
Direct coupled transistor logic	DCTL	
Direct current	dc, (DC-preferred)	
Direct current working volts	dcwv, DCWV	
Direction finding	D/F	
Distance	———	d
Double pole, double throw	DPDT	
Double pole, single throw	DPST	
Electric displacement	———	D
Electromotive force	emf, EMF	
Electronic Industries Association	EIA	
Electron volt	———	eV
Emitter coupled logic	ECL	
Energy-work	———	J
Equation	eq, EQ	
External	ext, EXT	

MEANING	ABBREVIATION	SYMBOL
Farad (unit of capacitance)	———	F
Field effect transistor	FET	
Filament	Fil	
Foot	FT	ft
Foot per second	FPS	ft/s
Foot poundal	ft-pdl	
Foot-pound force	FT LB F	ft . lb
Foot-second	ft-sec	
Force	———	N
Frequency	F, Freq	
Frequency (angular)	———	ω
Frequency modulation	fm, (FM preferred)	
Gain	———	A
Gauss (unit of flux density)	———	β
Giga (1 x 10^{+9})	———	G
Giga electron volt	———	GeV
Gigahertz	———	GHz
Gilbert (old unit for magnetomotive force)	gb, mmf	
Gram	G	g
Gravity	g	
Greenwich Mean Time (civil time)	GMT, GCT	
Ground	gnd, GND	
Hecto (1 x 10^{+2} normally avoided)	h	
Henry (unit of inductance)	———	H
Hertz (not per second)	———	Hz
High frequency	hf, HF	
Horsepower	HP	hp
Hour	HR	h
Impedance	———	Z
Inch	IN	in
Inch per second	IPS	in/s
Inductance (mutual)	———	M
Inductance (self)	———	L
Inductive reactance	———	X_L
Infrared	IR	
Inside diameter	ID	
Institute of Electrical and Electronic Engineers	IEEE	
Insulated gate field effect transistor	IGFET	
Integrated circuit	IC	
Intensity-magnetic field	———	H
Intermediate frequency	if, (IF-preferred)	
International Electrotechnical Commission	IEC	
International Standards Organization	ISO	
Interrupted continuous wave	icw	

MEANING	ABBREVIATION	SYMBOL
Joint electron device engineering council	JEDEC	
Junction diode	cr, CR	
Kilo ($1x10^{+3}$)	k	
Kilogram	——	kg
Kilohertz	——	kHz
Kilohm	(slang usage-K)	k Ω
Kilometre (preferred spelling)	——	km
Kilovar	kvar	
Kilovolt	(slang usage-K volts)	kV
Kilovolt ampere	——	kVA
Kilowatt	——	kW
Kilowatt hour	——	kWh
Lambert	L	
Large scale integrated circuit	LSI	
Length	——	l, L
Light emitting diode	LED	
Load resistor	——	R$_L$
Logarithm	Log	
Low frequency	LF	
Magnetic flux density	——	β
Magnetic flux (Maxwell)	——	ϕ, Φ
Magnetic field strength	——	H
Magnetomotive force	mmf	
Mass	——	m. M
Maximum	MAX	
Maxwell	mx	
Medium scale integrated circuit	MSI	
Mega ($1x10^{+6}$)	m, M	M
Megahertz	——	MHz
Megavolt	MV	
Megohm	Meg (slang)	M Ω
Metal oxide semiconductor	MOS	
Metal oxide semiconductor field effect transistor	MOSFET	
Metre* (measurement of length)	——	m
MHO	——	A/V, MHO
Micro ($1x10^{-6}$)	——	μ
Microampre	——	μA
Microfarad	mfd, MFD, μfd	μF
Microhenry	——	μH
Micrometre (preferred spelling)	——	μm
Micromho (replaced by microsiemen μs)	——	μmho, S
Micromicrofarad (replaced by picofarad pF)	——	$\mu\mu$fd, mmfd, (pF preferred)
Microvolt	micro V, μV	

*Metre or meter; first spelling is preferred.

MEANING	ABBREVIATION	SYMBOL
Microwatt	micro W, μW	
Milihenry	———	mH
Military	mil	
Milli (1×10^{-3})	———	m
Milliampere	———	mA
Millibar	———	m bar
Milligram	———	mg
Millimetre (preferred spelling)	———	mm
Millimho	———	m mho
Millions of cycles	(old) mega cps	(new) MHz
Millisecond	———	ms
Millivolt	———	mV
Milliwatt	———	mW
Minimum	MIN	
Minute	MIN	min
Modulated continuous wave	MCW	
Modulation, amplitude	am, AM	AM
Modulation factor	———	M
Modulation, frequency	fm, FM	FM
Modulation, percent of	———	M
Modulation, phase	pm, PM	PM
Modulation, pulse amplitude	pam, PAM	PAM
Modulation, pulse code	pcm, PCM	PCM
Modulation, pulse duration	pdm, PDM	PDM
Modulation, pulse position	ppm, PPM	PPM
Modulation, pulse width	pwm, PWM	PWM
Mutual conductance	———	gm, GM
Mutual inductance	———	M
Nano (1×10^{-9})	———	n
Nanoampere	———	nA
Nanofarad	———	nF
Nanohenry	———	nH
Nanometre (preferred spelling)	———	nm
Nanosecond	———	ns
Nanowatt	———	nW
National Electrical Manufactures Association	NEMA	
Negative	neg, NEG	B-
Neutralizing capacitor	C_n	
Newton	N	
No - connection	NC	
Normally closed	NC	
Normally open	NO	
Number	no, #	
Oersted (obsolete unit of magnetic field intensity)	———	H

MEANING	ABBREVIATION	SYMBOL
Ohm (unit of resistance)	———	Ω
Ohmmeter	———	Ωm
Ounce	oz	(slang usage)
Outside diameter	OD	
Peak	P, PK	
Peak inverse voltage	PIV	
Peak reverse voltage	PRV	
Peak to peak	PP, PK-PK	
Period	———	T
Permeability (relative/absolute)	———	μ
Permeance	———	P
Phase locked loop	PLL	
PI	———	π
Pico (1×10^{-12})	p	
Picoampere	———	pA
Picofarad	(old) mmf, MMF, $\mu\mu$f	pF (preferred)
Picosecond	pS	
Picowatt	pW	
Positive	pos, POS	B+
Potentiometer	pot, POT	R
Pound	LB	lb
Pound per square foot	PSF	lb/ft²
Pound per square inch	PSI	lb/in²
Power	P	
Power (average)	Pav	
Power factor	PF	
Power (peak)	Ppk	
Pressure	pres, P	
Primary	PRI	
Pulse amplitude modulation	PAM	PAM
Pulse code modulation	PCM	PCM
Pulse duration modulation	PDM	PDM
Pulse position modulation	PPM	PPM
Pulse recurrence frequency	PRF	
Pulse recurrence time	PRT	
Pulse repetition frequency	PRF	
Pulse width	PW	
Pulse width modulation	PWM	PWM
Quality factor (XL to R ratio)	———	Q
Quantity of electricity (coil or capacitor merit)	———	Q
Quiescent point	Q point	Q
Quiet automatic volume control	QAVC	
Radian	———	rad
Radio Electronics Television Manufacturers Association	RETMA	

Radio frequency	rf, RF	RF
Reactance	———	X
Receiver	rcvr	
Reference	ref	
Reluctance	rel	*R*
Resistance	res	R
Resistivity	———	P
Resistor transistor logic	RTL	
Resonance frequency	———	Fr
Revolutions per minute	RPM	r/min
Revolutions per second	RPS	r/s
Root mean square	rms, RMS	
Secant	sec	
Second (time)	SEC	s
Secondary	sec	
Sensitivity	SENS	
Single cotton covered (wire covering)	SCC	
Single cotton enamel (wire covering)	SCE	
Single pole, double throw switch	SPDT	
Single pole, single throw switch	SPST	
Single sideband	SSB	
Single silk covered (wire covering)	SSC	
Silicon controlled rectifier	SCR	
Sine	sin	
Square	SQ	
Square foot	SQ FT	ft^2
Square inch	SQ IN	in^2
Square metre (preferred spelling)	———	m^2
Standard	STD	
Standing wave ratio	SWR	
Super high frequency	SHF	
Susceptance	———	β
Switch	SW	
Synchronous	SYNC	
Tangent	TAN	
Television	TV	
Temperature (absolute)	temp	T
Temperature coefficient	———	Tc
Temperature (degrees centigrade)	———	°C
Temperature (degrees fahrenheit)	———	°F
Tera ($1x10^{+12}$)	T	
Terahertz	THz	
Thickness	———	t
Thermistor	Rt	t
Thousand of cycles/hertz per second	KC, KHz	
Time	———	t
Time constant	———	TC

MEANING	ABBREVIATION	SYMBOL
Time of one cycle/hertz (dependent on frequency)	———	T
Transformer	xformer,	T
Transmit-receive	TR	
Transistor-transistor logic	TTL	
Transistorized voltmeter	TVM	
Traveling wave tube	TWT	
Tuned radio frequency	TRF	
Turns (number)	———	N
Ultra high frequency	uhf, UHF	
Ultraviolet	UV	
Unijunction transistor	UJT	
United States of America Standards Institute	USASI	
Vacuum tube	V	
Vacuum tube voltmeter	VTVM	
Variable frequency oscillator	VFO	
Velocity of light (radiowaves)	———	C
Versus	VS, vs	
Very high frequency	vhf, VHF	
Very low frequency	vlf, VLF	
Voltage	Volt	V
Voltampere	———	VA
Voltage (average value)	V_{av}	E_{av}
Voltage controlled oscillator	VCO	
Voltage (effective/rms value)	Vrms, Eeff	E
Voltage (instantaneous value)	Vinst	e
Voltage (maximum value)	Vmax	Emax
Voltage regulator	———	VR
Voltage standing wave ratio	VSWR	
Volt-ohm-milliameter	VOM	
Volume control	VC	
Volume unit	Vu	
Volume unit meter	Vu-m	
Watt	W	W, P
Watt hour	WH	Wh
Wavelength	———	λ
Weber	Wb	

1.5   MATHEMATICAL SIGNS AND SYMBOLS

Everyone of us who has worked at a particular job for any duration of time develops and uses short-cuts or tricks of the trade. In mathematics, the same thing is true. One of the more popular tricks of the trade is to use symbols or abbreviations that reduce unnecessary writing. The following signs and symbols are those which are often used in our mathematical processes.

SYMBOLS	MEANING
$+$	plus; addition; plus; "or" gate
$-$	minus; subtraction; negative
$\pm$	tolerance; plus or minus, positive or negative
$A \div B$, A/B, $\dfrac{A}{B}$, $A:B$	division; A divided by B
$A \times B$, $A \cdot B$, AB, $(A)(B)$	multiplication; A times B
$(\,)$, $\{\,\}$, $[\,]$	parentheses, braces and brackets for grouping
$=$, $::$	equal to
$\cong$, $\approx$	approximately equal to
$\neq$	not equal to
$\equiv$	identical to
$<$	less than
$\nless$	not less than
$\leqq$	equal or less than
$>$	greater than
$\geqq$	equal or greater than
$\ngtr$	not greater than
$::$, $:$	proportional to (equals)
$:$	ratio
$\longrightarrow$	approaches as a limit
∞	infinity
Δ	delta; increment of; change in
$\sqrt{\ }$, $\sqrt{\ }$	square root
$\sqrt[n]{\ }$	Nth root
X^n	exponent of X; Nth power of X
Log, Log_{10}	common logarithm
Ln, Log e	hyperbolic, natural or napierian logarithm
e, ϵ	epsilon; base of natural (hyperbolic) logarithm; 2.718
π	Pi; 3.1416
$\dfrac{1}{2\pi}$	0.159
$\angle$	angle
$\llcorner$	right angle
$\perp$, $\perp$	perpendicular to
$\parallel$	parallel to; parallel
$|N|$	absolute value of N
$\bar{N}$	average value of N
X^{-n}, $\dfrac{1}{X^n}$	reciprocal of Nth power of X
$N°$	N degrees
N'	N minutes; N feet
N''	N seconds; N inches

SYMBOLS	MEANING
$\therefore$	therefore
F(x)	Function of (X)
C, K	constant (when following an equal sign)
$\int$	integration (integral in calculus)
$_b\int^a$	integral between limits of a and b
α	varies directly as
!	factorial
Σ	summation
j	"j" operator, square root of minus one
ω	angular velocity $2\pi F$
Lim.	limit value of an expression
Sin	sine
Cos	cosine
Tan, tg, tang	tangent
Cot, ctg	cotangent
Sec	secant
Cosec	cosecant
Versin	versed sine
Covers	coversed sine

1.6   SOLID STATE SCHEMATIC SYMBOLS

<u>REGION OHMIC CONNECTIONS</u>

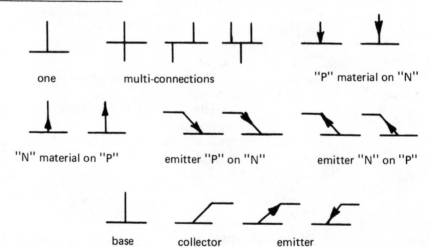

one   multi-connections   "P" material on "N"

"N" material on "P"   emitter "P" on "N"   emitter "N" on "P"

base   collector   emitter

<u>INTRINSIC REGION BETWEEN CONDUCTIVITY REGIONS</u>

dis-similar   collector   similar   collector

NOTES FOR SEMICONDUCTORS

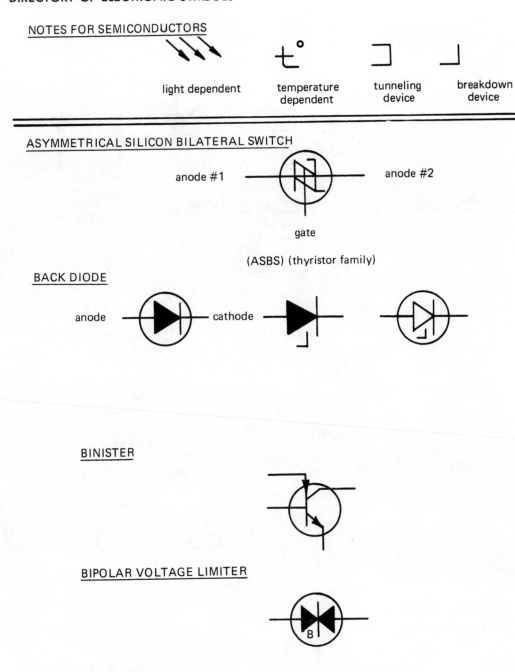

light dependent   temperature   tunneling   breakdown
                     dependent   device   device

ASYMMETRICAL SILICON BILATERAL SWITCH

anode #1   anode #2

gate

(ASBS) (thyristor family)

BACK DIODE

anode   cathode

BINISTER

BIPOLAR VOLTAGE LIMITER

BREAKDOWN DIODE
(bidirectional family)

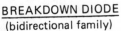
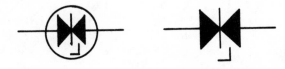

CAPACITIVE DIODE
(varactor family)

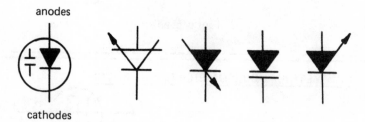

anodes

cathodes

COMPLEMENTARY SILICON CONTROLLED RECTIFIER

anode

gate

cathode

(CSCR)

COMPLEMENTARY UNIJUNCTION TRANSISTOR

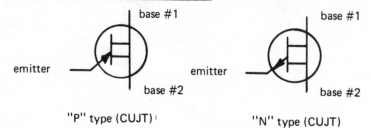

base #1

emitter

base #2

"P" type (CUJT)

base #1

emitter

base #2

"N" type (CUJT)

DARLINGTON AMPLIFIER

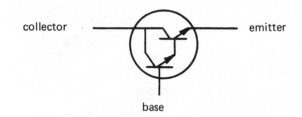

collector

emitter

base

DIAC
(thyristor family)

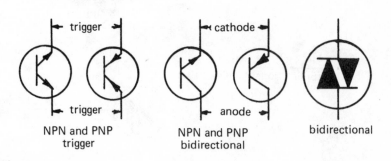

trigger

trigger

NPN and PNP
trigger

cathode

anode

NPN and PNP
bidirectional

bidirectional

DIODE/RECTIFIER

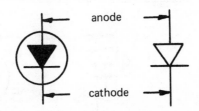

FIELD EFFECT TRANSISTOR (FET)

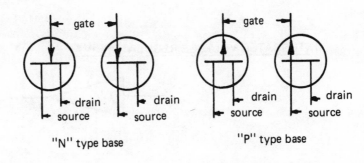

FIELD EFFECT TRANSISTOR—INSULATED GATE FIELD EFFECT (IGFET)

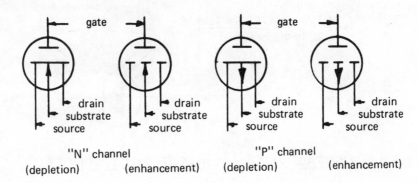

INSULATED GATE FIELD EFFECT TRANSISTOR
(MOSFET family)

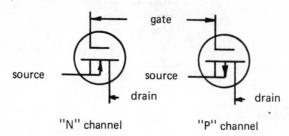

INTEGRATED VOLTAGE REGULATOR (IVA)

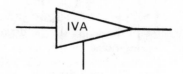

LIGHT ACTIVATED PROGRAMMABLE UNIJUNCTION TRANSISTOR
(LAPUT) (thyristor family)

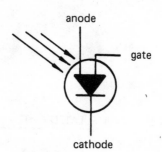

LIGHT ACTIVATED SWITCH
(LAS) (thyristor family)

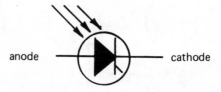

LIGHT ACTIVATED SILICON CONTROLLED RECTIFIER
(LASCR) (thyristor family)

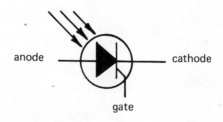

LIGHT ACTIVATED SILICON CONTROLLED SWITCH
(LASCS) (thyristor family)

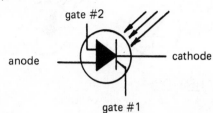

LIGHT EMITTING DIODE (LED)

LIGHT SENSITIVE DARLINGTON AMPLIFIER

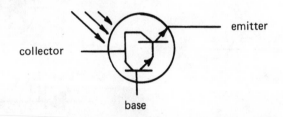

METAL OXIDE SEMICONDUCTOR FIELD EFFECT TRANSISTOR
(MOSFET) (thyristor family)

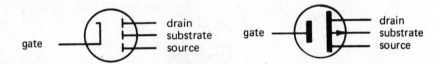

NPNP TRIODE SWITCH

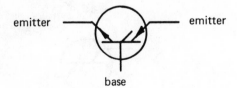

NPN TRANSISTOR

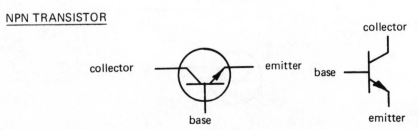

PHOTO DIODE

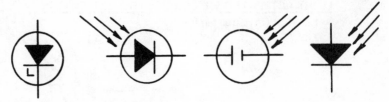

PHOTO TRANSISTOR

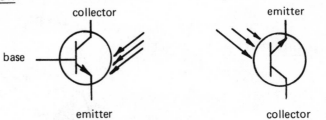

PIN DIODE

PIN TRIODE

PNP TRANSISTOR

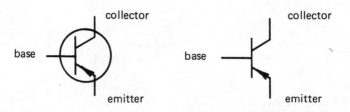

PNPN DIODE

PNPN TRIODE SWITCH

PROGRAMMABLE UNIJUNCTION TRANSISTOR
(PUT) (thyristor family)

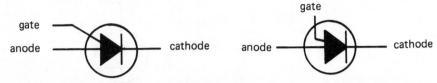

SHOCKLEY DIODE

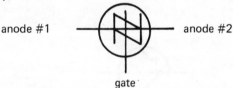

SILICON BILATERAL SWITCH
(SBS) (thyristor family)

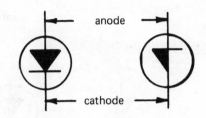

SILICON CONTROLLED RECTIFIER
(SCR) (thyristor family)

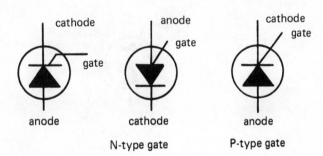

SILICON CONTROLLED SWITCH
(SCS) (thyristor family)

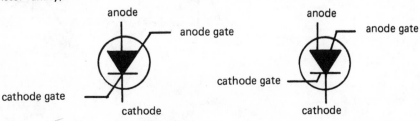

SILICON UNILATERAL SWITCH
(SUS) (thyristor family)

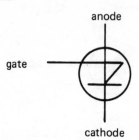

TEMPERATURE SENSITIVE DIODE

TETRODE TRANSISTOR

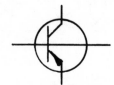

THYRECTOR

TRIAC

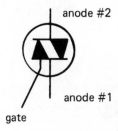 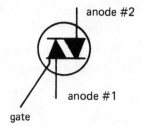

TRIGISTOR

TRIODE (BIDIRECTIONAL)

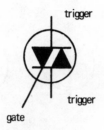

TUNNEL DIODE

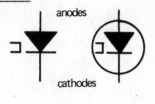 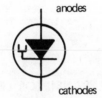

UNIJUNCTION TRANSISTOR (UJT)

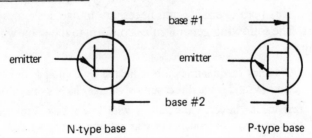

N-type base   P-type base

VARACTOR

ZENER DIODE (ALSO REFERRED TO AS FOLLOWS)

    1. BACKWARD DIODE
    2. BREAKDOWN DIODE
    3. AVALANCHE DIODE
    4. VOLTAGE REGULATOR

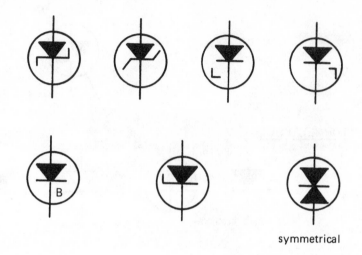

symmetrical

1.7   SIGNIFICANCE OF TRANSISTOR LETTER SYMBOLS

Since the development of solid state technology has increased the number of abbreviations, quantity letter subscripts, and symbols often encountered in the field of Electronics, it is both desirable and necessary to use some system to designate circuit parameters. The following general rules conform to the somewhat standardized practice used today.

1. DC values of quantities are indicated by capital letters with capital subscripts (V_{CE}), while direct supply voltages have subscripts repeated (V_{CC}).

2. rms values have capital letters with lower case subscripts (V_{cb}), while time-varying components of currents and voltages have lower-case letters with lower-case subscripts, (v_{cb}).

3. Instantaneous values are indicated with lower-case letters with upper-case subscripts (v_{BE}).

4. Maximum or peak values are like rms values but have an additional subscript "m" (E_{bem}).

5. Circuit elements are designated with capital letters (R_{10}) and transistor parameters have lower-case symbols (h_{11}).

1.7-1   Transistor Letter Symbols for Current

MEANING	LETTER SYMBOL
Average output rectified current	Io
Base current (instantaneous-AC)	ib
Changing base current (AC-rms)	Ib
Changing collector current (AC)	ic
Collector current (DC*-between base and emitter)	ICEX * specified circuit
Collector current (rms)	Ic
Collector cutoff DC current (base open)	ICEO
Collector cutoff DC current (base short circuited to emitter)	ICES
Collector cutoff DC current (emmiter open)	ICBO
Collector cutoff DC current (specified resistance between base and emitter)	ICER
Collector cutoff DC current (specified voltage between base and emitter)	ICEV
Common base forward current ratio (large signal, short circuit)	HFb
Common emitter forward current ratio (large signal, short circuit)	HFe
Common collector forward current ratio (large signal, short circuit)	HFc
Current amplification with output short circuited	H21
Current gain	Ai
Current stability factor	SI
DC base current	IB
DC collector current	IC
Drain current (DC)	ID
Drain cutoff current	ID(off)
Drain current-zero gate voltage	IDSS
DC emitter current (total)	Ie
Emitter DC current	IE
Emitter-collector offset current	IEC(OFS)
Emitter current (instantaneous)	ie
Emitter cutoff DC current (collector open)	IEBO
Emitter cutoff DC current (base short circuited to collector)	IECS
Forward current (alternating component)	If
Forward current (DC)	IF
Forward current (DC with alternating component)	IF(AV)
Forward current (instantaneous)	iF
Forward current (overload)	IF(OV)
Forward current (peak surge)	IFSM
Forward current (peak total value)	IFM
Forward current transfer ratio for common base (static value)	hFB

MEANING	LETTER SYMBOL
Forward current transfer ratio for common base (small signal, short circuit)	h_{fb}
Forward current transfer ratio for common collector (static value)	h_{FC}
Forward current transfer ratio for common collector (small signal, short circuit)	h_{fc}
Forward current transfer ratio for common emitter (static value)	h_{FE}
Forward current transfer ratio for common emitter (small signal, short circuit)	h_{fe}
Gate current (DC)	I_G
Gate current (forward)	I_{GF}
Gate current (reverse)	I_{GR}
Peak point current (double-base transistors)	I_P
Regulator current (DC reference)	I_Z
Regulator current-reference current (DC max rated)	I_{ZM}
Regulator current-reference current (DC near breakdown knee)	I_{ZK}
Reverse current (alternating components-rms)	I_r
Reverse current (instantaneous)	i_r
Reverse current (total-rms)	i_R (rms)
Reverse DC current	I_R
Reverse recovery current	i_R (REC)
Source current (zero gate voltage)	I_{SDS}
Valley point current (double base transistors)	I_V

1.7-2   Transistor Letter Symbols for Voltages

MEANING	LETTER SYMBOL
Base-collector DC voltage	V_{BC}
Base-collector voltage (rms)	V_{bc}
Base-collector voltage (instantaneous)	v_{bc}
Base-emitter voltage (rms)	V_{be}
Base supply DC voltage	V_{BB}
Breakdown reverse voltage	BV_R
Breakdown voltage between collector-base (emitter open)	BV_{CBO}
Breakdown voltage between collector-emitter (base open)	BV_{CEO}
Breakdown voltage between collector-emitter (base short circuited to emitter)	BV_{CES}
Breakdown voltage between emitter-base (collector open)	BV_{EBO}
Changing voltage (ac) between base-collector	V_{bc}
Changing voltage (ac) between base-emitter	V_{be}

MEANING	LETTER SYMBOL
Changing voltage (ac) between collector-emitter	V_{ce}
Collector DC supply voltage	V_{CC}
Collector to base voltage (rms)	V_{cb}
Collector to base voltage (instantaneous)	v_{cb}
Collector to emitter saturation voltage	$V_{ce\,(sat)}$
Collector to emitter voltage (instantaneous)	v_{ce}
Emitter to base DC voltage	V_{EB}
Emitter to base voltage (rms)	V_{eb}
Emitter to base voltage (instantaneous)	v_{eb}
Emitter to collector DC voltage	V_{EC}
Emitter to collector voltage (rms)	V_{ec}
Emitter to collector voltage (instantaneous)	v_{ec}
Emitter supply DC voltage	V_{EE}
Fixed base-emitter DC voltage	V_{BE}
Fixed collector-emitter DC voltage	V_{CE}
Forward DC voltage	V_F
Forward voltage (instantaneous)	v_f
Input voltage	$E\ in$
Output voltage	$E\ out$
Reach through voltage	V_{RT}
Reverse DC voltage	V_R
Reverse voltage (instantaneous)	v_r
Reverse voltage transfer ratio for common base (small signal, open circuit)	h_{rb}
Reverse voltage transfer ratio for common collector (small signal, open circuit)	h_{rc}
Reverse voltage transfer ratio for common emitter (small signal, open circuit)	h_{re}
Saturation voltage	$V\ sat$
Source voltage	V_g, V_s
Voltage feedback ratio with input open circuited	H_{12}
Voltage gain	A_v
Voltage stability factor	S_v

1.7-3 Transistor Resistances Letter Symbols

MEANING	LETTER SYMBOL
AC base resistance	r_b
AC collector resistance	r_c
AC emitter resistance	r_e
Collector-emitter saturation resistance	$r_{CE\,(sat)}$
Common base input resistance (static value)	h_{IB}
Common collector input resistance (static value)	h_{IC}
Common emitter intput resistance (static value)	h_{IE}
Forward transfer resistance with input open	r_{fe}

MEANING	LETTER SYMBOL
Input resistance of common base (output short circuited)	hib
Input resistance of common collector (output short circuited)	hic
Input resistance of common emitter (output short circuited)	hie
Input resistance with output open	rie
Load resistance	RL
Mutual resistance	rm
Output resistance with input open	roe
Reverse transfer resistance with input open	rre

1.7-4   Transistor Amplification Letter Symbols

MEANING	LETTER SYMBOL
Average power gain of common base (large signal)	GPB
Average power gain of common base (small signal)	Gpb
Average power gain of common collector (large signal)	GPC
Average power gain of common collector (small signal)	Gpc
Average power gain of common emitter (large signal)	GPE
Average power gain of common emitter (small signal)	Gpe
Forward short circuit current amplification factor (common base)	α fb
Forward short circuit current amplification factor (common collector)	α fc
Forward short circuit current amplification factor (common emitter)	α fe
Reverse open circuited voltage amplification factor (common base)	μ rb
Reverse common circuited voltage amplification factor (common collector)	μ rc
Reverse open circuited voltage amplification factor (common emitter)	μ re

1.7-5   Transistor Conductance Letter Symbols

Note: You will find the unit of siemens (s) replacing the unit of mhos (g)

MEANING	LETTER SYMBOL
Base-collector conductance	gbc
Base-emitter conductance	gbe
Collector-emitter conductance	gce

MEANING	LETTER SYMBOL
Common base output conductance (open circuit, static value)	h_{OB}
Common collector output conductance (open circuit, static value)	h_{OC}
Common emitter output conductance (open circuit, static value)	h_{OE}
Forward transfer conductance	g_{fe}
Input conductance	g_{ie}
Output conductance	g_{oe}
Output conductance of common base with input open	h_{ob}
Output conductance of common collector with input open	h_{oc}
Output conductance of common emitter with input open	h_{oe}
Reverse transfer conductance	g_{re}

1.7-6 Transistor Capitance Letter Symbols

MEANING	LETTER SYMBOL
Common base input capacitance	C_{ib}
Common base input capacitance (open circuit)	C_{ibo}
Common base input capacitance (short circuit)	C_{ibs}
Common base output capacitance	C_{ob}
Common base output capacitance (open circuit)	C_{obo}
Common base output capacitance (short circuit)	C_{obs}
Common base reverse transfer capacitance short circuit)	C_{rbs}
Common collector input capacitance	C_{ic}
Common collector output capitance	C_{oc}
Common collector reverse transfer capacitance (short circuit)	C_{rcs}
Common emitter intput capacitance	C_{ie}
Common emitter intput capacitance (open circuit)	C_{ieo}
Common emitter input capacitance (short circuit)	C_{ies}
Common emitter output capacitance	C_{oe}
Common emitter output capacitance (open circuit)	C_{oeo}
Common emitter output capacitance (short circuit)	C_{oes}
Common emitter reverse transfer capacitance (short circuit)	C_{res}
Drain source capacitance (gate shorted to source)	C_{oss}
Gate source capacitance (drain shorted to source)	C_{iss}
Interelement capacitance (base-collector)	C_{cb}
Interelement capacitance (base-emitter)	C_{eb}
Interelement capacitance (collector-emitter)	C_{ce}
Reflected capacitance	C_{sp}

1.7-7   Transistor Power Letter Symbols

MEANING	LETTER SYMBOL
Base-emitter total power input (DC, average)	P_{BE}
Base-emitter total power input (instantaneous)	p_{BE}
Collector-base total power input (DC, average)	P_{CB}
Collector-base total power input (instantaneous)	p_{CB}
Collector-emitter total power input (DC, average)	P_{CE}
Collector-emitter total power input (instantaneous)	p_{CE}
Common base input power (large signal)	P_{IB}
Common base input power (small signal)	P_{ib}
Common base output power (large signal)	P_{OB}
Common base output power (small signal)	P_{ob}
Common collector input power (large signal)	P_{IC}
Common collector input power (small signal)	P_{ic}
Common collector output power (large signal)	P_{OC}
Common collector output power (small signal)	P_{oc}
Common emitter input power (large signal)	P_{IE}
Common emitter input power (small signal)	P_{ie}
Common emitter output power (large signal)	P_{OE}
Common emitter output power (small signal)	P_{oe}
Emitter-base total power input (DC, average)	P_{EB}
Emitter-base total power intput (instantaneous)	p_{ES}

1.7-8   Miscellaneous Transistor Letter Symbols

MEANING	LETTER SYMBOL
Ambient temperature	T_A
Base electrode	B, b
Case temperature	T_C
Collector electrode	C, c
Combination of "N" type and "P" type semiconductor	PN
Common base configuration	CB
Common collector configuration	CC
Common emitter configuration	CE
Common emitter input impedance (small signal, short circuited)	h_{ie}
Common base static transconductance	g_{Mb}
Common emitter static transconductance	g_{Me}
Common collector static transconductance	g_{Mc}
Conversion loss	L_c
Cutoff frequency	f_α
Emitter electrode	E, e
Fall time	t_f
Forward recovery time	t_{fr}
Hybrid	h

MEANING	LETTER SYMBOL
Imput impedance with output short circuited	H_{11}
Junction temperature	T_j or T_J
Large signal breakdown impedance	BZ
Large signal transconductance (common base)	G_{mb}
Large signal transconductance (common collector)	G_{mc}
Large signal transconductance (common emitter)	G_{me}
Noise figure	NF
Operating temperature	T_{opr}
Output admittance with input open circuited	H_{22}
Pulse average time	t_w
Pulse time	t_p
Reverse recovery time	t_{rr}
Rise time	t_r
Semiconductor with acceptor impurity	P type
Semiconductor with donor impurity	N type
Small signal breakdown impedance	b_z
Small signal transconductance (common base)	g_{mb}
Small signal transconductance (common collector)	g_{mc}
Small signal transconductance (common emitter)	g_{me}
Storage temperature	T_{stg}
Storage time	t_s
Temperature	T
Time delay	t_d
Total power input (DC, average)	P_T
Total power input (instantaneous)	p_T
Transistor with one "N" type and two "P" type semi-conductors	PNP
Transistor with one "P" type and Two "N" type semi-conductors	NPN

1.8   SIGNIFICANCE OF VACUUM TUBE LETTERS

The standardized notation system for the vacuum tube proposed by the Institute of Radio Engineers has lessened the probable confusion that may have existed when defining voltages and currents. Because of this system, we now know whether the specified voltage or current is the supply, the value between the grid and cathode, with a signal or without a signal applied, peak, rms, or instantaneous value. The following is included as a handy reference for those of us who have forgotten just what represents and means what.

All capital E's and I's designate voltage and currents for static conditions; they refer to the peak, rms, or DC values. All small e's and i's designate the instantaneous values. Subscripts b, c and l refer to total voltages between designated tube element points and the cathode. The subscripts g, p, and z indicate the alternating components present at these same corresponding reference points. Figure 1.1 illustrates typical circuitry applications.

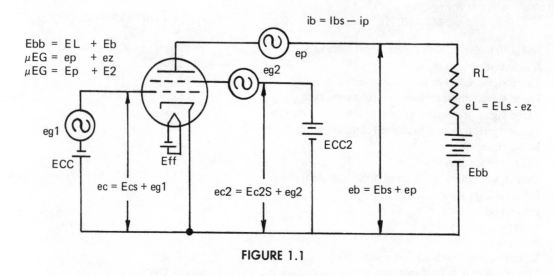

FIGURE 1.1

1.8-1   Vacuum Tube Electrode Letters

MEANING	LETTER SYMBOL
Cathode	K
Control grid	C_1, G, G_1
Heater/filament	H, F
Plate/anode	P
Screen grid	C_2, G_2
Suppressor grid	C_3, G_3

1.8-2   Vacuum Tube Letters for Currents

MEANING	LETTER SYMBOL
Average control grid current (with signal)	I_{cs}
Average load current (with signal)	I_{ls}
Average plate current (with signal)	$\underline{I}_{bs}$
Average/quiescent value of grid current	$\underline{I}_{gl}$
Average/quiescent value of plate current	I_p
Average screen grid current (with signal)	I_{g2s}
Alternating control grid current (rms value)	I_g
Alternating control grid current (peak value)	I_{gm}
Alternating control grid current (instantaneous value)	i_g
Alternating load current (instantaneous value)	i_z
Alternating load current (peak value)	I_{zm}
Alternating load current (rms value)	I_z
Alternating plate current (instantaneous value)	i_p
Alternating plate current (peak value)	I_{pm}
Alternating plate current (rms value)	I_p

MEANING   LETTER SYMBOL

Alternating screen grid current (instantaneous value)	I_{g2}
Alternating screen grid current (peak value)	I_{g2m}
Alternating screen grid current (rms value)	I_{g2}
Control grid current (no signal)	I_c
Filament/heater current	I_f
Instantaneous control grid current	i_c
Instantaneous load current	i_l
Instantaneous plate current	i_b
Instantaneous screen grid current	i_{c2}
Load current (no signal)	I_l
Plate current (no signal)	I_b
Screen grid current	I_{c2}

1.8-3   Vacuum Tube Letters for Voltages

MEANING   LETTER SYMBOL

Average control grid to cathode voltage (with signal)	E_{cs}
Average control grid voltage	E_c
Average plate to cathode voltage (with signal)	E_{bs}
Average plate voltage	E_b
Average screen grid to cathode voltage (with signal)	E_{c2s}
Average voltage drop across load (with signal)	E_{ls}
Alternating component across load (instantaneous value)	e_z
Alternating component on control grid (instantaneous value)	e_g
Alternating component on plate (instantaneous value)	e_p
Alternating component on screen grid (instantaneous value)	e_{g2}
Alternating voltage across load (peak value)	E_{zm}
Alternating voltage across load (rms value)	E_z
Alternating voltage on control grid (peak value)	E_{gm}
Alternating voltage on control grid (rms value)	E_g
Alternating voltage on plate (peak value)	E_{pm}
Alternating voltage on plate (rms value)	E_p
Alternating voltage on screen grid (peak value)	E_{g2m}
Alternating voltage on screen grid (rms value)	E_{g2}
Control grid supply voltage	E_{cc}
Control grid to cathode voltage (no signal)	E_c
Heater/filament supply voltage	E_{ff}, E_f
Instantaneous control grid to cathode voltage	e_c
Instantaneous plate to cathode voltage/ instantaneous total plate voltage	e_b

MEANING	LETTER SYMBOL
Instantaneous screen grid to cathode voltage	e_{c2}
Instantaneous voltage across load	e_L
Plate supply voltage	E_{bb}, B, B+
	E source
Plate to cathode voltage (no signal)	E_b
Screen grid supply voltage	E_{cc2}
Screen grid to cathode voltage (no signal)	E_{c2}
Voltage drop across load (no signal)	E_L

1.8-4   Vacuum Tube Resistance Letters

MEANING	LETTER SYMBOL
Plate resistance (AC)	r_p
Plate resistance (DC)	R_P

1.9   VACUUM TUBE SCHEMATIC SYMBOLS

BEAM FORMING PLATE ELECTRODES

BEAM POWER AMPLIFIER

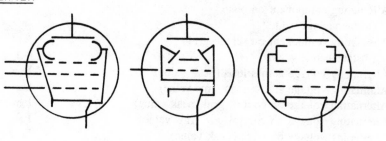

CATHODE ELECTRODE

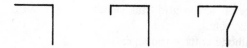

CATHODE RAY INDICATOR

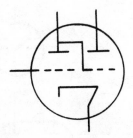

CATHODE RAY TUBE

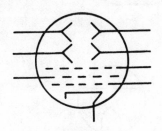

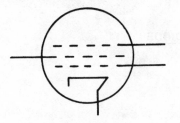

electrostatic deflection   magnetic deflection

COLD CATHODE ELECTRODE

COLD CATHODE GAS DIODE

DEFLECTING ELECTRODE

DIODE

directly heated   indirectly heated

DOUBLE CAVITY ENVELOPE

DUAL TRIODE

DUO-DIODE

DUO-DIODE TRIODE

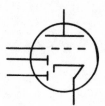

DYNODE ELECTRODE

EYE TUBE

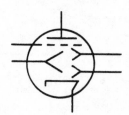

EXCITER ELECTRODE

EXCITRON WITH GRID AND HOLDING ANODE

(mercury pool tube)

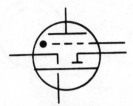

FILAMENT/HEATER ELECTRODE

GAS FILLED TUBE

GLASS ENVELOPE

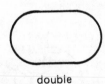

single   double   split

GRID ELECTRODE

——— – – – –

IGNITER (MERCURY POOL TUBE)

IGNITRON WITH GRID

(mercury pool tube)

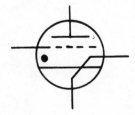

INTERNAL TUBE SHIELDING

LOOP COUPLING ELECTRODE

MAGNETRON

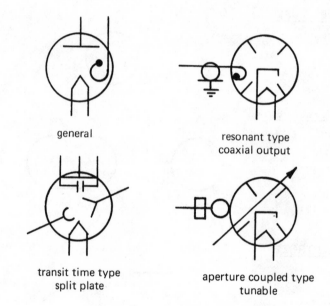

general

resonant type
coaxial output

transit time type
split plate

aperture coupled type
tunable

MULTIPLIER PHOTO TUBE

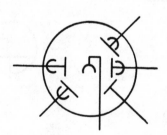

PENTAGRID CONVERTER

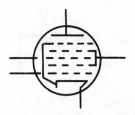

PENTODE

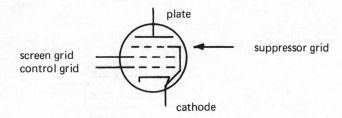

plate

screen grid
control grid

suppressor grid

cathode

PHOTO ELECTRIC CATHODE ELECTRODE

PHOTO TUBE

PLATE ELECTRODE

POOL CATHODE ELECTRODE

SINGLE CAVITY ENVELOPE

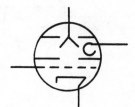

SPLIT MAGNETRON

TETRODE

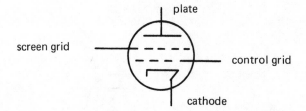

plate

screen grid

control grid

cathode

TRAVELING WAVE TUBE

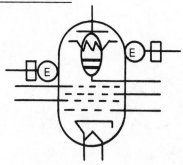

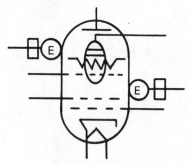

forward wave amplifier

backward wave amplifier

TRANSMIT-RECEIVE TUBE

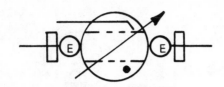

tunable aperture coupled

TRIODE

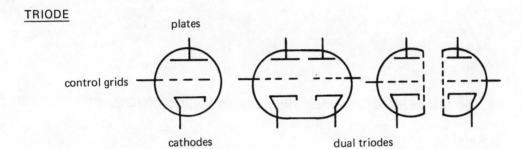

plates

control grids

cathodes   dual triodes

VELOCITY MODULATED TUBE

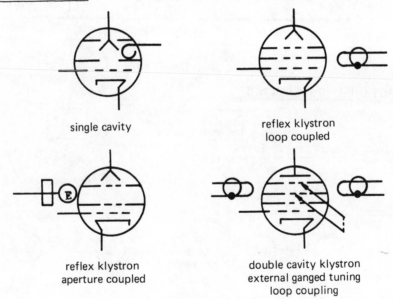

single cavity

reflex klystron
loop coupled

reflex klystron
aperture coupled

double cavity klystron
external ganged tuning
loop coupling

X-RAY TARGET ELECTRODE

<u>X-RAY TUBE</u>

filamentary cathode
with focusing grid

electrostatic
shielding

accelerating electrode
(control grid)

1.10   ARCHITECTURAL ELECTRICAL/ELECTRONIC SYMBOLS

<u>BUS DUCTS/WIRE WAYS</u>

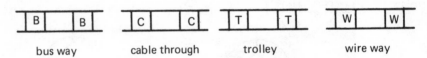

bus way   cable through   trolley   wire way

<u>DISTRIBUTION (UNDERGROUND)</u>

hand hole   man hole

<u>LIGHTING OUTLETS</u>

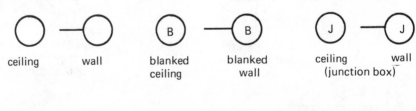

ceiling   wall   blanked   blanked   ceiling   wall
                          ceiling   wall   (junction box)

exit-ceiling exit-wall
(surface/suspended)

fluorescent
(surface)
(suspended)

fluorescent
(continuous row)

fluorescent-strip
(bare lamp)

LIGHTING OUTLETS-RECESSED

| ceiling | wall | exit-ceiling | exit-wall | fluorescent |

PANEL BOARD/SWITCH BOARD

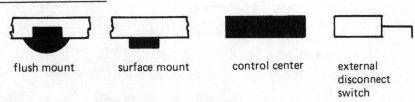

| flush mount | surface mount | control center | external disconnect switch |

RECEPTACLE OUTLETS

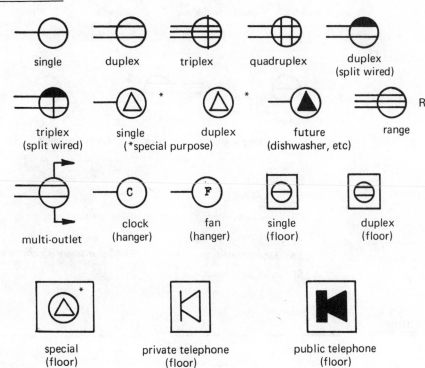

single   duplex   triplex   quadruplex   duplex (split wired)

triplex (split wired)   single (*special purpose)   duplex   future (dishwasher, etc)   range   R

multi-outlet   clock (hanger)   fan (hanger)   single (floor)   duplex (floor)

special (floor)   private telephone (floor)   public telephone (floor)

NOTE: GROUNDED OUTLETS HAVE A "G" NEXT TO THE SAME SYMBOLS.

SIGNALING OUTLETS (INDUSTRIAL-COMMERCIAL-INSTITUTIONS)

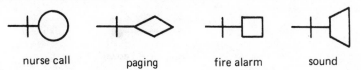

| nurse call | paging | fire alarm | sound |

SIGNALING OUTLETS (RESIDENTIAL)

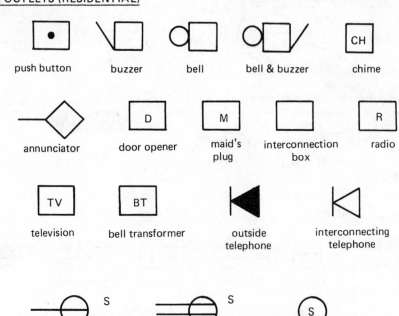

| push button | buzzer | bell | bell & buzzer | chime |

| annunciator | door opener | maid's plug | interconnection box | radio |

| television | bell transformer | outside telephone | interconnecting telephone |

SWITCHES

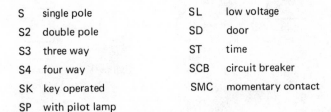

| single | double | ceiling pull |

NOTE: LETTERS WITH SWITCHES HAVE THE FOLLOWING MEANING:

S	single pole	SL	low voltage
S2	double pole	SD	door
S3	three way	ST	time
S4	four way	SCB	circuit breaker
SK	key operated	SMC	momentary contact
SP	with pilot lamp		

WIRING

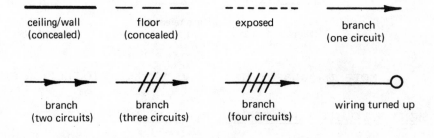

| ceiling/wall (concealed) | floor (concealed) | exposed | branch (one circuit) |

| branch (two circuits) | branch (three circuits) | branch (four circuits) | wiring turned up |

wiring turned down

CHAPTER 2

Essential Mathematical Data
for Electronic Application

In order for you to work effectively in any facet of electronics, you are expected to use and understand mathematical concepts. Some of these concepts include: arithmetic, algebra, geometry and trigonometry; along with these are complex numbers, vectors, logarithms, decibels, binary and octonary numbers, graphs, Boolean algebra, and advanced mathematics, etc.

This chapter includes the type of data you will need to know for mastering electronics mathematics. Factual concepts will be written in almost a shorthand style, thereby avoiding a storybook approach and possibly some of those things which many of us call "nice to know" facts.

A general set of rules for mathematical calculation which provides a correct order for operations is listed here to aid in organizing your thoughts before any data is presented.

1. First add and subtract numbers, any numbers contained within brackets or parentheses that are to be divided, multiplied, raised to a power, or factored.

2. Multiply and divide numbers whose product or quotient is to be raised to a power or factored.

3. Determine the value for powers and roots.

4. Multiply and divide numbers within the terms.

5. Add and subtract the terms.

2.1   ARITHMETIC

2.1-1   Addition

Addition is the process used to find total quantity of any like thing represented by two or more numbers. For the sake of terminology, the number to be increased by addition of another number is called the *augend*, while the number to be added to the augend is called the *addend*. Obviously, the end result is the total, termed the *sum*.

Numbers can be added correctly only if they have related values. For example:

10 amperes (augend)		10 volts
+ 5 amperes (addend)	NOT	+ 5 amperes
15 amperes (sum)		15 but meaningless units

The magnitude of the units must be expressed in the same terms! You can add 10 amperes and 1000 milliamperes only if you convert one value into the other expressed value. This must be done although both are values expressing specific amounts of current.

$$
\begin{array}{lll}
10 & \text{amperes} & \text{Change to} \qquad 10 \text{ amperes} \\
+1000 & \text{milliamperes} & \qquad\qquad\quad\; +\; 1 \text{ ampere} \\
& & \qquad\qquad\quad\; \overline{11 \text{ amperes}}
\end{array}
$$

Oftentimes electronic calculations are written in such the same manner as is the writing on this line of print you are reading. The plus symbol indicates the addition process and is used to connect the terms. For example:

$$
\begin{array}{lll}
1,000 \text{ mA} & + \; 10,000 \text{ mA} & = \; 11,000 \text{ mA} \\
\text{1st term} & + \; \text{2nd terms} & = \; \text{sum}
\end{array}
$$

2.1-2   Subtraction

The operation involved in determining the difference between two like-numbered units is called subtraction. The minus symbol separates the terms when written on one line sequence. If you had $XL = 45$ ohms and $XC = 30$ ohms the reactance value would be:

$$
\begin{array}{lll}
45 \text{ ohms} \; - \; 30 \text{ ohms} = 15 \text{ ohms} & \text{OR} & 45 \text{ ohms} \\
(XL) \; - \; (XC) = (Z) & & -30 \text{ ohms} \\
& & \overline{15 \text{ ohms}}
\end{array}
$$

The number 45 is called the *minuend,* the value of 30 represents the *subtrahend* and the number of 15 is termed the *remainder*. As in the addition process, make sure all numbers represent the same unit of electronic measurement, that is keep kilowatts with kilowatts, and microfarads with microfarads, convert if necessary.

2.1-3   Multiplication

Multiplication is indicated whenever a times sign ($\times$) is inserted between numbers, or when a dot ($\cdot$) is seen, or when parentheses ( ) or brackets [ ] are used or when letter or letter-number combinations are written as AB, or 4A, etc. The number to be multiplied is called the *multiplicand,* the number by which the mutliplicand is multiplied by is the *multiplier* and the answer or result is termed the *product*.

If three resistors in a series circuit had the same resistance of 10 ohms, the total resistance could be obtained by adding the 10 ohm value (concrete value) together three times to achieve the answer. We could, however, multiply the concrete value of 10 ohms by a no measurable unit value of 3 to obtain the same answer having the same related units of ohms as the concrete number.

Electronics utilizes the facts which when multiplied together, even though their units of measure aren't the same, result in success. For example, volts times amperage equals watts, amperage times resistance equals voltage. Each has its own unit of measurement, but the *product* is in the desired unit.

2.1-4   Division

The indicated operation for division might be seen: written (ten divided by 5), or $\frac{10}{5}$, or 10 ÷ 5, or 10/5, 5$\lfloor$10 or 5$\lceil$10, all of these mean the same. The number to be divided (10) is the *dividend,* the *divisor* (5) is the number being divided into the dividend, and the answer is called the *quotient.*

Numbers having the same or different values of measurable units can be divided and still obtain the desired quotient. For example, the total resistance of a series circuit is 1000 ohms and each of the ten resistors has the same resistance: what then is the value of each resistor? Solution: divide the 1000 units of total resistance by the non measurable unit 10, indicating the number of resistors. The answer 100 is still in the related unit of ohms. The same units might be illustrated via amplifier output voltage gain where output volts divided by input volts equals a number having no units. A different unit illustration is solving for resistance when voltage and current values are known.

Resistance in ohms equals voltage in volts divided by current in amperes. The answer is in the unit of ohms.

2.1-5   Fractions

All fractions have some number divided by another number. The number written above the horizontal line (divide symbol) or to the left of a diagonal line (divide symbol) is called the *numerator* and the number under the horizontal line or to the right of the diagonal line is called the *denominator*. Since fractions actually denote division, all of the rules for division apply. In reality, the numerator is the dividend and the denominator is the divisor.

Fractions are called *simple* if the numerator and denominator are whole numbers called *integers*. A *complex* fraction is one which has either a numerator or denominator in fraction form. *Proper* fractions are those which have smaller numerators than denominators. *Improper* fractions are those whose numerators are equal to or greater than their denominators. The term *mixed number* refers to a number plus a fraction combination.

There may be many times when working with fractions that you will need to alter the numerator and denominator of a fraction without changing its value. The reasons normally for doing this are:

1. Simplifying by reducing to its lowest terms.

2. To find a common denominator for addition or subtraction operations.

3. To change improper fractions to mixed numbers for sake of recognition.

4. To change mixed numbers to improper fractions for multiplication or division operations.

Alteration without change is possible if the numerator and denominator are multiplied or divided by the same number.

Hence:

$$\frac{1}{5} \times \frac{1}{1} = \frac{1}{5}$$

OR

$$\frac{1}{5} \times \frac{2}{2} = \frac{2}{10} \qquad \text{AND} \qquad \frac{2}{10} = \frac{1}{5}$$

OR

$$\frac{1}{5} \times \frac{8}{8} = \frac{8}{40} \qquad \text{AND} \qquad \frac{8}{40} = \frac{1}{5}$$

All of these equal each other.

In changing improper fractions to mixed numbers, perform the division process as indicated. For example:

$$\frac{19}{3} = \begin{array}{r} 6 \\ 3\overline{)19} \\ 18 \\ \hline 1 \end{array} = 6\tfrac{1}{3}$$

In changing mixed numbers to improper fractions follow this guide. For example: $8^{10}/_{16}$

1. Change the whole number 8 into an equivalent number.

$$8 \times \frac{16}{16} = \frac{128}{16}$$

2. Add the fraction to the whole number fractional equivalent.

$$\frac{128}{16} + \frac{10}{16} = \frac{138}{16}$$

Note: Mixed numbers must be changed to improper fractions before division or multiplication processes.

2.1-5A ADDITION AND SUBTRACTION OF FRACTIONS

There are two basic rules to follow when adding fractions.

Rule 1: when fractions are added, their denominators must be the same; if they are different, you must convert to a common denominator. For example:

$$\frac{3}{8} + \frac{3}{4} + \frac{1}{2} = \frac{3}{8} + \frac{6}{8} + \frac{4}{8}$$

Rule 2: After having a common denominator, add the numerators but keep the same common denominator. the solution, using the previous example, is:

$$\frac{3}{8} + \frac{6}{8} + \frac{4}{8} = \frac{13}{8} \quad \text{or } 1\tfrac{5}{8}$$

There are also two basic rules when subtracting fractions.

Rule 1: is the same as for adding; if the denominators are different you must convert to a common denominator. For example:

$$\frac{10}{8} - \frac{1}{4} - \frac{1}{2} = \frac{10}{8} - \frac{2}{8} - \frac{4}{8}$$

Rule 2: After obtaining the common denominator, subtract the numerators, but keep the denominator. The answer is:

$$\frac{10}{8} - \frac{2}{8} - \frac{4}{8} = \frac{10-2-4}{8} = \frac{10-6}{8} = \frac{4}{8} \text{ OR } \frac{1}{2}$$

2.1-5B   MULTIPLICATION AND DIVISION OF FRACTIONS

When multiplying a whole number (integer) by a fraction or vice versa, multiply the numerator of the fraction by the whole number and write this product over the denominator. In other words, divide the denominator into the new numerator. For example:

$$\frac{3}{16} \times 3 = \frac{9}{16}$$

To multiply one fraction by another fraction, you have to first multiply the numerators together, then you multiply the denominators together; if they aren't the same, then divide the new denominator into the new numerator. To illustrate this:

$$\frac{5}{3} \times \frac{2}{4} = \frac{10}{12} \text{ OR } \frac{5}{6}$$

When dividing a fraction by a whole number, consider the whole number as a fraction since one is its denominator, then invert the divisor and multiply. For example:

$$\frac{3}{8} \div 4 = \frac{3}{8} \div \frac{4}{1} = \frac{3}{8} \times \frac{1}{4} = \frac{3}{32}$$

To divide a whole number by a fraction the same basic concept holds true because:

$$10 \div \frac{4}{5} = \frac{10}{1} \times \frac{5}{4} = \frac{50}{4}$$

or a fraction by a fraction:

$$\frac{9}{16} \div \frac{1}{8} = \frac{9}{16} \times \frac{8}{1} = \frac{72}{16}$$

Note: When dividing mixed numbers, first change the mixed numbers to improper fractions then proceed as previously indicated.

2.1-6   Powers and Roots

Oftentimes mathematics used in Electronics can seem confusing, but in reality if you understand the signs and symbols, the only real problem is cranking out the answers. *Powers* of numbers are denoted by an *exponent* written above and to the right of the number. This is a simplified way of writing that a number is multiplied by itself a number of times. An example would be solving for power when the current is 5 amps and the resistance is 10 ohms. Rather than writing $P=5\times5\times10$, it is written $P=(5)^2 \times 10$, which means the same thing $(P=I^2R)$.

Finding *roots* of a number is simply the reverse of finding the power. The radical sign and the exponent are the two notations used to designate the root of a number. If a radical sign ($\sqrt{}$) does not have a number called an *index* with it, the square root operation is indicated. If, however, an index number is assigned with the radical sign, then that root operation is indicated. For example: The square root of 81 is written $\sqrt{81}$, the cube root of 27 is $\sqrt[3]{27}$, the fifth root of 243 is $\sqrt[5]{243}$, etc. Exponent notation is almost as simple because square roots are illustrated by the exponent ½, cube roots by the exponent ⅓, fifth roots by the exponent $^1/_5$, etc.

To re-illustrate this method the previous examples will be written as $81^{½}$ (square root of 81), $27^{⅓}$ (cube root of 27) and $243^{1/_5}$ for the fifth root of 243.

2.1-6A   SQUARE ROOT CALCULATIONS

Many occasions arise in Electronics calculations when square roots must be found and your handy square root table is elsewhere. These general rules will revive your memory for your square root task.

Rule 1.   Divide the number into two digit groups beginning from the decimal point.

Rule 2.   Find the largest number whose value when squared is equal to or less than the digits in the first number group.

Rule 3.   Write this number above the first group.

Rule 4.   Write the square of that number under the first group, and subtract.

Rule 5.   Draw a division line above the remainder found after subtracting, then write the digits found in the second group of numbers to the right of the remainder. The resulting number is called the dividend.

This next step involves a multiple guess type of approach because a trial divisor will be discovered.

Rule 6.   Multiply the first digit of the root (first number written in rule 3) by two and write this product to the left of the dividend. This number, when multiplied by 10, is the trial divisor.

Rule 7.   Determine how many times the trial divisor can be divided into the dividend. remember it must be equal to or less than the dividend. Write this number above the second group of numbers

Rule 8.   Add this new number, called the second digit of the root, to the trial divisor, then mutliply the complete divisor by the second digit of the root and write this product under the dividend.

Note: If the product is greater than the dividend, the complete trial divisor and second digit of the root must be decreased by one and rules 7 and 8 must be repeated.

Rule 9.   Subtract the product from the dividend, write the third group of numbers to the right of the remainder to complete a new dividend. Draw a division line above this dividend.

Rule 10. Multiply the first two digits of the root by two, write this product to the left of the dividend.

Rule 11. Find out how many times the new divisor, when mutliplied by ten, can be divided into the new dividend, then write this number above the third group of numbers.

Rule 12. Add this number to the trial divisor, then multiply the complete trial divisor by the third digit root number recorded above the third group of numbers, and then write this product under the dividend.

The next step is to repeat rules 9-12 as often as you wish, depending upon how many significant figures you wish. If you add zeroes, add them in groups of two thereby minimizing the confusion.

For example, we will show the square root for the number 243.

Rule #1 $\sqrt{2\ 43.}$

example rewritten: $\overset{1}{\sqrt{2\ 43.}}$

Rules #2 and 3 1

Rules #4 $\overline{\left.\right|1.43}$

Rules #5

example rewritten: $\overset{1}{\sqrt{2\ 43.}}$

Rule #6 $20\ \overline{\left.\right|1\ 43}^{\,1}$

example rewritten: $\overset{1\ 5}{\sqrt{2\ 43.00}}$

Rules #7 & 8 $25\ \overline{\left.\right|1\ 43}^{\,1}$

Rules #9 & 10 $30\ \overline{\left.\right|\ \ 18\ 00}^{\,1\ 25}$

example rewritten: $\overset{1\ 5.\ 5\ \ \ 8}{\sqrt{2\ 43.\ 00\ 00}}$

$25\ \overline{\left.\right|1\ 43}^{\,1}$

$305\ \overline{\left.\right|\ \ 18\ 00}^{\,1\ 25}$

$\overline{\left.\right|\ \ 2\ 75\ 00}^{\,15\ 25}$

Rules # 11 & 12 $3108\ \overline{\left.\right|\ \ \ \ 26\ 36}^{\,2\ 48\ 64}$

2.1-6B   ADDITION AND SUBTRACTION OF POWERS AND ROOTS

Most of us have discovered there is our way of doing something, the wrong way, and the right way. The right way when adding or subtracting powers and roots will provide you with the correct answer if you follow these rules and don't make any calculation errors.

Rule 1. Find the power or root value before adding or subtracting.

For example: To add 2^4 to $\sqrt{9}$ and subtract 10, first determine the power and root values as shown:

$$\sqrt{9} + \quad 2^4 - 10 =$$
$$3 + 16 \;\; - 10 =$$
$$19 \;\; - 10 = 9$$

2.1-6C   MULTIPLICATION AND DIVISION OF POWERS AND ROOTS

Multiplication or division processes can be performed before finding root or power equivalents for a number if all of the numbers having the power or root designations are of equal value. *The exponents are added during multiplication processes as indicated.*

Multiplication example for powers
$$10 \times 10^2 \times 10^3 = 10^{1+2+3} = 10^6$$

Multiplication example for roots
$$\sqrt{3} \times \sqrt{3} = 3^{\frac{1}{2}}$$
$$3^{\frac{1}{2}} = 3^{2/2} = 3$$

Multiplication example for power and roots
$$3^2 \times \sqrt{3} = 3^{4/2} \times 3^{\frac{1}{2}} = 3^{4/2 + \frac{1}{2}} = 3^{5/2}$$

What $3\frac{5}{2}$ really means $\sqrt{3^5}$ or the square root of the number three, multiplied by itself five times. This could be solved using two general methods: the first is by multiplying $3 \times 3 \times \sqrt{3}$ and obtaining 9×1.73 or 15.57 as the answer; another method would be to multiply three by itself five times, obtaining the number 243 and then take the square root of 243. The final answer would 15.58.

Note: If a number expressed as a power or root is raised to another power or factored into another root, the exponents are mutliplied. The number three squared raised to the third power is written $(3^2)^3$ and is equal to 3^6. The square root of the number five cubed is written $(5^3)^{\frac{1}{2}}$. If you follow the above suggestion, $5^{3/2}$ would be its equivalent.

All of the previous information involving multiplying powers of roots had the same number as its base. If the numbers having power or root designations are unequal, you must determine the power and root of the number before mutliplying. To illustrate this procedure we will multiply five squared by four cubed. First determine what five squared equals then what four cubed equals and then multiply.

$$5^2 \times 4^3 = 25 \times 64 = 1600$$

Dividing like numbers having exponents is as simple as multiplying if you know the rule when performing the division process. The rule for dividing like numbers with exponents is: subtract the exponent of the divisor from the exponent of the dividend. The following examples illustrate this.

$$3^5 \div 3^3 = 3^{5-3} = 3^2$$

OR

$$\frac{3^5}{3^3} = \frac{3 \times 3 \times 3 \times 3 \times 3}{3 \times 3 \times 3} = 3 \times 3$$

2.1-7   Ratios

A *ratio* is a simple means of comparing two quantities which are measured in the same unit. The gain of an amplifier circuit is commonly expressed as:

$$gain = \frac{\text{output voltage}}{\text{input voltage}}$$

If the input signal is 0.3 volts and the output signal is 18 volts the gain ratio would be:

$$gain = \frac{18v}{0.3v} = \frac{60}{1} = 60$$

OR

$$gain = 18:0.3 = 60:1$$

For what it's worth, when written in ratio form (60:1), the number 60 is called the *antecedent* and the number 1 is called the *consequent*. The antecedent is the same as dividend and the consequent is the same as denominator or divisor. Regardless of what you call them, both terms of the ratio may be multiplied or divided by the same number without changing the value of the ratio.

2.1-8   Proportions

A proportion is actually an equality of ratios. For example, a series circuit forms a voltage divider and if 100 volts are applied to the input and 10 volts appeared across the output, then the proportions could be written in any of the following ways:

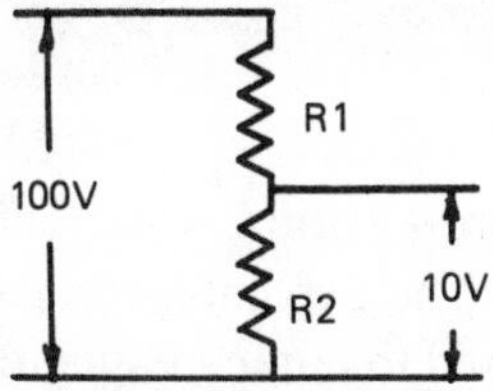

Figure 2.1

10:100::R2: (R1 + R2)

OR

10:100 = R2: (R 1 + R 2)

OR

$$\frac{10}{100} = \frac{R2}{R1 + R2}$$

The first term (10) and the fourth term (R1 + R2) are called *extremes*. The second term (100) and third term (R2) are called *means*.

In order to find the value of one unknown term and to prove a proportion is correct, any one of the following rules or methods may be used. For simplicity, the previous example will be used except that now resistor R1 will be 18 ohms and resistor R2 will be 2 ohms.

Rule 1: Product of the means equals the product of the extremes.

$$10{:}100 = R2 \, (R1 + R2)$$

$$10{:}100 = 2 : 20$$

product of means $100 \times 2 = 200$

product of extremes $10 \times 20 = 200$

Rule 2: Product of means divided by either of the two extremes equals the other extreme.

$$10{:}100 = 2 : 20$$

product of means $100 \times 2 = 200$

$$\frac{200}{10} = 20 \text{ OR } \frac{200}{20} = 10$$

Rule 3: Product of extremes divided by either of the two means equals the other means.

$$10{:}100 = 2{:}20$$

product of extremes $10 \times 20 = 200$

$$\frac{200}{100} = 2 \text{ OR } \frac{200}{2} = 100$$

2.1-9   Inverse Proportions

Two numbers are inversely proportional when one number will increase in value while the other number is decreased in value.

Ohms Law is a good illustration of inverse proportions in action because Ohms Law states that the current (I) through a resistor (R) is equal to the voltage (E) dropped across the resistor divided by the resistor's resistance. Needless to say, if the voltage is constant and the resistance is doubled, the current is halved or if the voltage remains the same and the resistance is halved, the current is doubled. The current flow through the resistor is inversely proportional to the resistance.

2.1-10   Percentages

Percentages in electronics are used to express proportions, efficiencies, changes, and tolerances of various components and circuits. The percentage notation, represented by the symbol %, means by the hundred and is used to compare quantities. The percentage range is from zero to 100%.

To find a quantity expressed as a percentage of a number, convert the percent to a decimal or common fractional value and mutliply the number by this value. For example:

60% off list price of 5.85 is:
5.85 × 60% = 5.85 × 0.6 = 3.57

An application of finding the percentage of one quantity with respect to another quantity would be expressing efficiency. If an amplifier has an input wattage of 110 and has an out wattage rating of 50, its percentage of efficiency is easily found by dividing the output by the input, move the decimal point two places to the right, and then add the percent symbol. For example

$$\text{Efficiency} = \frac{\text{output power}}{\text{input power}}$$

$$\text{Efficiency} = \frac{50}{110} \times 100\% = 0.4545 \times 100\% = 45.45\%$$

2.1-10A CONVERTING DECIMALS TO PERCENT

The percent notation is considered to represent two decimal places since 1% is equal to 0.01 in decimal form. Any decimal can be converted to a percent by moving the decimal two places to the right and by adding the percent symbol. For example 0.159 equal 15.9%.

2.1-10B CONVERTING COMMON FRACTIONS TO PERCENT

Whenever converting a common fraction to a percent, you must first change the fraction to a decimal by dividing the numerator by the denominator, and then convert the decimal to a percent. The following example illustrates this process:

$$\frac{1}{8} = 0.125 \times 100\% = 12.5\%$$

2.1-10C CONVERTING PERCENT TO DECIMALS

Omit the percent symbol and move the decimal point two places to the left to change a percent to a decimal. The following equation shows this conversion:

$$63.6\% = 0.636$$

2.1-10D CONVERTING PERCENT TO COMMON FRACTIONS

A number written as a percent is converted to a common fraction by omitting the percent symbol, writing the number which had the percent symbol as a numerator and 100 as the denominator. This forms the fraction. Reduce it into its lowest terms as shown, then convert to fractional equivalents.

$$37.5\% = \frac{37.5}{100} = 0.375 = \frac{3}{8}$$

2.1-11   Scientific Notation

A thorough knowledge of the powers to ten and their use will greatly assist you in solving problems. Often it is necessary to make calculations involving certain electronic values, and both large and small numbers are used. Use of the powers of ten will enable you to work problems having very small decimal values or very large decimal whole numbers with minimum difficulty.

Since we manipulate large and small numbers using powers of ten; they are listed here through the twelfth power. Only twelve are illustrated because seldom is there need to go beyond this exponent in electronics.

POWERS OF TEN

$10^0 = 1$	$10^0 = 1$
$10^1 = 10$	$10^{-1} = 0.1$
$10^2 = 100$	$10^{-2} = 0.01$
$10^3 = 1000$	$10^{-3} = 0.001$
$10^4 = 10,000$	$10^{-4} = 0.0001$
$10^5 = 100,000$	$10^{-5} = 0.000,01$
$10^6 = 1,000,000$	$10^{-6} = 0.000,001$
$10^7 = 10,000,000$	$10^{-7} = 0.000,000,1$
$10^8 = 100,000,000$	$10^{-8} = 0.000,000,01$
$10^9 = 1,000,000,000$	$10^{-9} = 0.000,000,001$
$10^{10} = 10,000,000,000$	$10^{-10} = 0.000,000,000,1$
$10^{11} = 100,000,000,000$	$10^{-11} = 0.000,000,000,01$
$10^{12} = 1,000,000,000,000$	$10^{-12} = 0.000,000,000,001$

Any decimal fraction may be readily expressed by the scientific notation method (the number ten times a negative power of ten) while multiples of ten expressed as ten to the proper positive power of ten represent large decimal values.

For convenience in calculations the numbers to the left of the decimal point are limited to a one-digit number (a number between one and ten), and the remainder of the significant figures are placed to the right of the decimal point. The power of ten multiplier is adjusted accordingly. For example, various fractions and whole numbers containing the digits 3142 may be expressed as follows:

$$0.0003142 = 3.142 \times 10^{-4}$$
$$0.003142 = 3.142 \times 10^{-3}$$
$$0.03142 = 3.142 \times 10^{-2}$$
$$0.3142 = 3.142 \times 10^{-1}$$
$$3.142 = 3.142 \times 10^{0}$$
$$31.42 = 3.142 \times 10^{1}$$
$$314.2 = 3.142 \times 10^{2}$$
$$3142.0 = 3.142 \times 10^{3}$$
$$31420.0 = 3.142 \times 10^{4}$$

Changing numbers to powers of ten

$$1.\ 0.000193 = 1.93 \times 10^{-4}$$

$$\begin{aligned}
&2.\ 0.0196 && = 1.96 \times 10^{-2} \\
&3.\ 177,560 && = 1.7756 \times 10^{5} \\
&4.\ 1,535 && = 1.535 \times 10^{3} \\
&5.\ 159,000 && = 1.59 \times 10^{5}
\end{aligned}$$

Changing powers of ten to decimals

$$\begin{aligned}
&1.\ 5.1 \times 10^{6} = 5,100,000. \\
&2.\ 3.5 \times 10^{3} = 3,500 \\
&3.\ 5.8 \times 10^{2} = 580 \\
&4.\ 1.60 \times 10^{-4} = 0.000,160 \\
&5.\ 6.28 \times 10^{-2} = 0.0628
\end{aligned}$$

2.1-11A   MULTIPLYING POWERS OF TEN

In multiplication; exponents having the same base (which is the present case) are added. Multiplication then is expressed as:

$$10^{a} \times 10^{b} = 10^{a+b}$$

For example, multiply 0.0027 by 135.8

Solution: (1st) convert both numbers to powers of ten

$$\begin{aligned}
0.0027 &= 2.7 \times 10^{-3} \\
135.8 &= 1.358 \times 10^{2}
\end{aligned}$$

(2nd) multiply the numbers and add the exponents.

$$3.6666 \times 10^{-1}$$

Note: Rules for multiplication of exponents must be employed.

2.1-11B   DIVIDING POWERS OF TEN

Division of numbers having exponents to the same base (10 in our case) is accomplished by subtracting the divisor exponent from the dividend exponent. Division is expressed as follows:

$$\frac{10^{a}}{10^{b}} = 10^{a-b}$$

For example divide 0.036666 by 0.00882

Solution: (1st) convert both numbers to powers of ten

$$\frac{0.036666}{0.00882} = \frac{3.6666 \times 10^{-2}}{8.82 \times 10^{-3}}$$

(2nd) divide the numbers and subtract the exponents

$$0.4157 \times 10^{1}$$

(3rd) convert to standard answer)

4.157

Note: Rules for dividing exponents must be employed.

2.1-11C   RAISING POWERS USING POWERS OF TEN

When raising a number expressed as a power of ten to an additional power, the exponent of the base is multiplied by the exponent denoting the power. This is expressed as:

$$(10^a)^b = 10^{ab}$$

for example, square the number 0.001

Solution: express the number as a power of ten

Since: $0.001 = 1 \times 10^{-3}$

Then $(1 \times 10^{-3})^2 = 1 \times 10^{-6} = 0.000001$

2.1-11D   EXTRACTING ROOTS USING POWERS OF TEN

When extracting roots of a number expressed as a power of ten, the exponent of the base is divided by the exponent indicating the root. This is expressed as:

$$\sqrt[b]{10^a} = 10^{a/b}$$

For example extract the square root of 10,000.

Solution: Write the number as a power of ten, then proceed.

$10,000 = 1 \times 10^4$

$\sqrt{10^4} = 10^{4/2} = 10^2 = 100$

2.2   ALGEBRA

Algebra is a more useful tool, when analyzing and solving technical problems, than is arithmetic. The main reason for its usefulness is because algebra has a unique combination of arithmetic, mathematics and alphabetical letters that are substituted for numbers. The common practice of using the first letters in the alphabet to represent the known quantities and the last of the letters in the alphabet to represent the unknown quantities has been modified somewhat for electronic calculations. Modification is necessary to avoid further confusion since specific electronic quantities are easily represented by using the beginning letter or the common electronic symbol. Some of these specific quantities are represented in this way: Power P, Impedance Z, Voltage E, Current I, Resistance R, etc.

These letters or symbols are called *literal numbers* or *general numbers*. Sometimes expressions may require individual recognition especially if several quantities use the same letter or symbol. *Subscripts* are used to indicate any differences and help provide identity for each quantity. These subscripts are actually small numbers or letters that are written somewhat below and to the right of the letter or symbol.

An algebraic *expression* is another way of saying *formula,* or since it indicates *equality* of quantities, an *equation*. Algebraic expressions are solved by substituting numerical values for the unknown symbols, and are reduced to their simplest form when performing the indicated operations. Addition, subtraction, multiplication, division and square roots used in algebra are handled in the same manner as they were used in arithmetic. Hints, laws and rules for each operation will be covered in later algebraic sections.

2.2-1   Coefficients and Terms

A number or letter written before a quantity (indicating multiplication) is called a *coefficient*. *Terms* are those numbers or groups of letters and numbers that are separated by a plus or minus sign. Algebraic expressions have their own identity based upon the number of terms an equation has. The *monomial* expression has only one term ($P = IE$) while the rest of the algebraic expressions are grouped under the general heading of being a *polynomial*. This simply means that the equation has more than one term. A *binomial* is a polynomial having two terms ($RT = R_1 + R_2$) and a polynomial having three terms is called a *trinomial* ($RT = R_1 + R_2 + R_3$).

All coefficients used in algebraic expressions consist of a sign being either positive or negative (written or understood) and a particular *magnitude*. The magnitude is called an *absolute value*. This number is or can be enclosed between two parallel vertical lines. For example, the absolute value of a 150 volt source regardless of its polarity is written $|150|$. The absolute value of -150 volts is the same as for 150 volts since the same amount of electrical force is indicated.

2.2-2   Addition

The following rules apply to algebraic addition processes:

1. When the numbers have like signs (both positive, or both negative) find the sum of the absolute values and then prefix the total with the common sign.
$$+8 +4 = +12 \quad \text{or} \quad -8 +(-4) = -12$$

2. If the numbers have unlike signs, determine the difference between the absolute values and then prefix the remainder with the sign of the larger absolute value.
$$+8 +(-4) = +4 \quad \text{or} \quad -8 +4 = -4$$

Note: If more than two terms are added, find the sum of the first two terms then add this sum to the third term, then add this second sum to the fourth term, etc.

2.2-3   Subtraction

The rule for algebraic subtraction involves multiplying the subtrahend by a minus one to change the subtrahend's sign. The problem is then that of addition since the algebraic sum is them found.

$$
\begin{array}{ccc}
+8 & +8 & \\
-(+4) = \underline{-4} & \text{OR} & -8 \quad -8 \\
+4 & & -(-4) = \underline{+4} \\
& & -4
\end{array}
$$

Note: If several terms are to be subtracted, change all the necessary signs required for subtractions, and then find the algebraic sum of the terms.

2.2-4   Multiplication

Algebraic multiplication is governed by the following rules:

1. The product of two numbers having like signs (either both positive or both negative) is positive.

$$+8 \times (+4) = +32 \quad \text{or} \quad -8 \times (-4) = +32$$

2. When multiplying two numbers having unlike signs, the product is negative.

$$-8 \times (+4) = -32 \quad \text{or} \quad +8 \times (-4) = -32$$

Note: To find the product of more than two numbers, multiply the first two numbers, then use that product to multiply by the third number, multiply that product by the fourth number, etc.

2.2-5   Division

The following rules apply to algebraic division operations:

1. Two numbers, divided by numbers having like signs, have a positive quotient.

$$\frac{+8}{+4} = +2 \quad \text{or} \quad \frac{-8}{-4} = +2$$

2. The quotient of two numbers having unlike signs will be negative.

$$\frac{+8}{-4} = -2 \quad \text{or} \quad \frac{-8}{+4} = -2$$

2.2-6   Powers/Exponents

A positive power is always obtained when a positive number is squared (multiplied by itself) and when a negative number is multiplied by itself an even number of times. For example:

1. $E \times E = E^2$
2. $(-3)^2 = (-3)(-3) = +9$
3. $E \times E \times E = E^3$
4. $(-2)^4 = (-2)(-2)(-2)(-2) = +16$

A negative power is obtained when negative numbers are raised to negative powers as indicated

$$(-2)^3 = (-2)(-2)(-2) = (-2)(+4) = -8$$
$$-E^3 = (-E)(-E)(-E)$$

Numbers raised to powers can be handled in the following two ways:

1st by Dividing $\quad \dfrac{3^2}{3^4} = \dfrac{3 \times 3}{3 \times 3 \times 3 \times 3} = \dfrac{1}{3^2}$

2nd. by Subtracting exponents $\quad \dfrac{3^2}{3^4} \qquad 3^{2-4} = 3^{-2}$

Any number having a negative power can be changed without altering its meaning in the manner following, by taking the reciprocal.

$$3^{-2} = \frac{1}{3^2}$$

2.2-7   Roots

A root is indicated by a radical sign and index number (unless a square root is understood), or by a fractional exponent which is the reciprocal of the index number. The general form for a root of a quantity may be written as $x\sqrt{A}$ or $A^{1/x}$. The letter A represents the quantity to be equally factored and is called the *radicand* when the radical sign is used and the *base* when the fractional exponent is used. The letter X represents the number of factors to be found. The rules for handling algebraic roots are as follows.

1. Odd roots of positive numbers are positive and odd roots of negative numbers are negative.

$$\sqrt[3]{27} = 3 \quad \text{or} \quad \sqrt[3]{-27} = -3$$

2. Even roots of a positive number can be either positive or negative and is often times indicated by the double positive-negative ($\pm$) sign.

$$\sqrt{9} = \pm\, 3$$

3. The root of a fraction is equal to the root of the numerator divided by the root of the denominator.

$$\sqrt{\frac{E^2}{R}} = \frac{\sqrt{E^2}}{\sqrt{R}} = \frac{E}{\sqrt{R}}$$

4. When finding the product of roots that have the same radicand, first write the roots in its fractional exponent form, and then write the product as a common base adding the exponents.

$$\sqrt{P} \times \sqrt[4]{P} = P^{1/2} \times P^{1/4} = P^{1/2 + 1/4} = P^{3/4}$$

5. To find quotients of roots having the same radicand, write the roots in their fractional exponent form, and then write the quotient as the common base to the exponent of the numerator minus the exponent of the denominator.

$$\frac{\sqrt{P}}{\sqrt[5]{P}} = P^{1/2} \times P^{-1/5} = P^{1/2 - 1/5} = P^{5/10 - 2/10} = P^{3/10}$$

6. *Rationalizing a denominator* is an operation including a fraction which has a square root quantity in its denominator. These fractions are changed into an integer when mutliplying the numerator and denominator by the square root quantity.

$$\frac{1}{\sqrt{2}} = \frac{1}{\sqrt{2}} \times \frac{\sqrt{2}}{\sqrt{2}} = \frac{\sqrt{2}}{\sqrt{2} \times \sqrt{2}} = \frac{\sqrt{2}}{2}$$

7. A square root in a two-term denominator of a fraction can be changed to an integer by multiplying both the numerator and denominator by the conjugate of the denominator. A conjugate is simply the sum of two terms and the difference of the same two terms. For example:

$$\frac{1}{4 + \sqrt{P}} = \frac{4 - \sqrt{P}}{(4 + \sqrt{P})(4 - \sqrt{P})} = \frac{4 - \sqrt{P}}{16 - \sqrt{P^2}} = \frac{4 - \sqrt{P}}{16 - P}$$

2.2-7A BASIC ALGEBRAIC ROOT RELATIONSHIPS

The following equations illustrate root identities that may prove their worth when solving problems.

1. $\sqrt[x]{AB} = \sqrt[x]{A} \cdot \sqrt[x]{B}$

2. $\sqrt{\dfrac{\sqrt[x]{A}}{B}} = \dfrac{\sqrt[x]{A}}{\sqrt[x]{B}}$

3. $\sqrt[x]{A} \cdot \sqrt[y]{A} = A^{\frac{1}{x} + \frac{1}{y}}$

4. $\dfrac{\sqrt[x]{A}}{\sqrt[y]{A}} = A^{\frac{1}{x} - \frac{1}{y}}$

5. $\dfrac{1}{\sqrt{A}} = \dfrac{\sqrt{A}}{A}$

6. $\left[\sqrt[x]{A}\right]^{\frac{1}{y}} = \sqrt[xy]{A}$

2.2-8 How to Handle Two-Term Mathematical Processes

Binomials are algebraic expressions consisting of two terms. These particular algebraic expressions are subject to mutliplication, division, addition, subtraction, raising to powers and taking roots. The following data will illustrate the basics necessary when accomplishing mathematical operations.

2.2-8A SQUARE OF BINOMIAL SUMS

The algebraic expression $A + B$, having two terms, called a binomal, can be squared by multiplying or by applying a simple rule.

$$
\begin{array}{r}
\text{Multiplying:} \quad A + B \\
A + B \\
\hline
A^2 + AB \\
+ AB + B^2 \\
\hline
A^2 + 2AB + B^2
\end{array}
$$

Rule: Square the first term, add twice the product of the two terms, then add the square of the second term:

$$(A + B)^2 = A^2 + 2AB + B^2$$

2.2-8B   SQUARE OF A BINOMIAL DIFFERENCE

The algebraic binomial expression $E = IR$ can be squared either by multiplying or by using a simple rule.

$$
\begin{array}{l}
\textit{Multiplying:}\quad E-IR \\
\qquad\qquad\quad\; E-IR \\
\qquad\qquad\;\; \overline{E^2-EIR} \\
\qquad\qquad\qquad -EIR + IR^2 \\
\qquad\qquad\;\; \overline{E^2-2EIR + IR^2}
\end{array}
$$

Rule: Square the first term, subtract twice the product of the two terms, then add the square of the second term.

$$(E-IR)^2 = E^2-2EIR + IR^2$$

2.2-8C   PRODUCT OF BINOMIAL SUMS AND DIFFERENCES

These two ways of solving algebraic expressions prove to be useful when handling complex numbers involving the "j" operator used in electronic calculations. To illustrate this $R + jX$ and $R-jX$ will be examined.

$$
\begin{array}{l}
\textit{Multiplication:}\quad R + jX \\
\qquad\qquad\qquad R- jX \\
\qquad\qquad\quad\; \overline{R^2 + RjX} \\
\qquad\qquad\qquad\quad - RjX - jX^2 \\
\qquad\qquad\quad\; \overline{R^2 \qquad\quad - jX^2}
\end{array}
$$

Rule: Square first term, then subtract the square of the second term.

$$(R + jX)(R-jX) = R^2 - jX^2$$

2.2-8D   BINOMIAL PRODUCTS HAVING ONE COMMON TERM

In the following illustrations the basic binomials will be $A + B$, $A + C$ for the multiplication of binomial sums; $A-B$, $A-C$ for the binomial differences, and $A + B$, $A-C$ for the binomial sum and differences.

Binomial sum
(multiplication)

$$
\begin{array}{l}
A + B \\
A + C \\
\overline{A^2 + AB} \\
\qquad\qquad + AC + BC \\
\overline{A^2 + AB + AC + BC}
\end{array}
$$

Binomial differences
(multiplication)

$$\begin{array}{r} A-B \\ A-C \\ \hline A^2 - AB \\ -AC + BC \\ \hline A^2 - AB - AC + BC \end{array}$$

Binomial sum and differences
(multiplication)

$$\begin{array}{r} A + B \\ A - C \\ \hline A^2 + AB \\ -AC - BC \\ \hline A^2 + AB - AC - BC \end{array}$$

2.2-9 Algebraic Fractions

Arithmetic operations involving addition, subtraction, division and multiplication of fractions are similar basic operations when working with algebraic fractions. The only real difference is that algebraic fractions may involve positive numbers, negative numbers, literal quantities and unknown quantities. To aid in algebraic fraction calculations the following will be given.

When working with fractions, three positions of the fraction must be examined for its implied mathematical sign. These three sign positions are: (1) directly preceding the whole fraction, (2) directly before the numerator, and (3) directly before the denominator.

Rule 1: Mathematical signs of any two of the three positions may be changed without altering the faction's value.

$$+ \frac{+A}{+B} = + \frac{-A}{-B} = - \frac{(-A)}{+B} = - \frac{+A}{(-B)}$$

Note: Polynomials can be changed also in fraction form by changing each sign for each term (MULTIPLYING BY A MINUS ONE).

$$Ax^2 - Bx + C = -(-Ax^2 + Bx - C)$$

2.2-9A ADDITION OR SUBTRACTION

Whenever adding or subtracting fractions, a common denominator must first be found. We will demonstrate only the addition process since the algebraic operation is the same. We will calculate the total current flowing through two resistors connected in parallel.

The formula is: $I_T = \dfrac{E R_1}{R_1} + \dfrac{E R_2}{R_2}$ **Figure 2-2**

Total current equals the voltage dropped across R_1 divided by the resistance of R_1 plus the voltage dropped across resistor 2 divided by the resistance of R_2.

Finding common denominator:

$$I_T = \frac{ER_1}{R_1} + \frac{ER_2}{R_2} = \frac{ER_1}{R_1R_2} + \frac{ER_2}{R_1\ R_2}$$

Rewrite equation by writing the algebraic sum of the numerators over the common denominator.

$$I_T = \frac{ER_1 + ER_2}{R_1R_2}$$

Since E is common to both terms in the numerator, it is factored out and thereby simplifies and solves the equation:

$$I_T = \frac{E(R_1 + R_2)}{R_1R_2}$$

2.2-9B   MULTIPLICATION

Algebraic and arithmetic fractions are multiplied by similar methods, in that multiplication of the numerators obtains the new numerator while mutliplying the denominators provides the new denominator. As in solving arithmetic fraction problems, equal factors found in both the numerator and denominator should be cancelled to simplify the operation. For example:

$$\frac{A}{B} \cdot \frac{C}{A} = \frac{AC}{BA} = \frac{C}{B}$$

2.2-9C   DIVISION

The same rules used in arithmetic when dividing fractions are used in algebraic fraction division. To divide by a fraction simply invert the divisor and multiply:

$$\frac{\frac{A}{B}}{\frac{C}{D}} = \frac{A}{B} \cdot \frac{D}{C} = \frac{AD}{BC}$$

2.2-10   Solving Algebraic Equations

Algebraic equations are easily solved when all of the rules involving algebra are employed as needed in the ways illustrated in this section of Chapter 2.

The general mathematical operations outlined under arithmetic in this chapter also have their place in solving algebraic equations as you may have already determined. All of this application of rules helps one to solve the equation for its unknown. Oftentimes in algebra, as in arithmetic, the equation must be reduced or changed into a simpler form. When this is done in algebra the process is called *factoring*.

2.2-10A   FACTORING

A *factor* is a common letter or number which appears in an equation. A *common factor* is that letter or number which appears in two or more terms in the same equation. The product of all factors that are common to a group of terms is called *"The Highest Common Factor."* An example for a common factor would be found in an equation used to find the source voltage (Es) for a series circuit having two resistors:

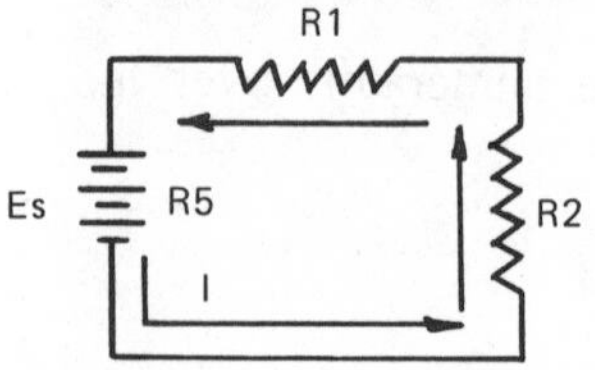

$$Es = IR_1 + IR_2$$

Figure 2-3

The terms on the right of the equal sign can be factored since the current (I) is the same through both resistors and therefore is common to both resistors $Es = IR_1 + IR_2$. The equation after factoring the common term would resemble:

$$Es = I(R_1 + R_2)$$

2.2-10B   SOLVING QUADRATIC EQUATIONS BY FACTORING

Whenever factoring the common factor from each term in an expression, arrange the resulting terms in an ascending or descending order to aid in determining whether it still can be factored. By doing this, you might discover a quadratic equation since every quadratic equation takes on the $AX^2 + BX + C = 0$ form. The letter X represents the unknown, and the letters A and B represent any positive or negative numerical value, including zero. In order to factor an equation, common factors must be determined. This simply means what times what equals the equation and what is common within the terms. To illustrate this the following equations show what expressions times what expressions are equal. (Refer to Section 2.2-8 for additional details).

$$(1) \quad A^2 + 2AB + B^2 = (A + B)^2$$
$$(2) \quad A^2 - 2AB + B^2 = (A - B)^2$$
$$(3) \quad A^2 - B^2 = (A + B)(A - B)$$

To factor this equation $X^2 + 9X + 14$, you must find the factors of 14 representing the C term in $AX^2 + BX + C$, paying particular attention to the values which when added together equals the numerical value of B in the $AX^2 + BX + C$ form.

The factors of 14 are:	*The sums of the factors are:*
$14 = 14 \times 1$	$14 + 1 = 15$
$14 = 2 \times 7$	$2 + 7 = 9$

Since the sum of the two factors 2 and 7 when added together equals the B term, these are the factors of the equation. The following equation may be written.

$$(X + 2)(X + 7) = X^2 + 9X + 14$$

To satisfy an equation the unknown or unknowns must be determined. In solving quadratic equations we arrange the equation so that it equals zero. When solving a quadratic equation having one unknown, arrange the equation so that one expression or member is the product of the two factors and the other member is zero. Set each of the factors equal to zero and solve for the unknown or unknowns. To illustrate this the previous example will be used.

1. Write the equation so the right-hand member equals zero.

$$X^2 + 9X + 14 = 0$$

2. Factor the left-hand member still equaling zero.

$$(X + 2)(X + 7) = 0$$

3. Solve each of the factors for the value of X when each factor is said to equal zero.

$$X + 2 = 0 \qquad X + 7 = 0$$
$$\text{AND}$$
$$X = -2 \qquad X = -7$$

This results in the roots of the equation, and when substituted for the unknown, will satisfy the original equation.

$$
\begin{array}{ll}
X^2 + 9X + 14 = 0 & \qquad X^2 + 9X + 14 = 0 \\
(-2)^2 + 9(-2) + 14 = 0 & \qquad (-7)^2 + 9(-7) + 14 = 0 \\
4 - 18 + 14 = 0 & \qquad 49 - 63 + 14 = 0 \\
-18 + 18 = 0 & \qquad -63 + 63 = 0 \\
0 = 0 & \qquad 0 = 0
\end{array}
$$

Since the equation is satisfied, you have found the correct roots.

2.2-10C   SOLVING QUADRATIC EQUATIONS BY COMPLETING THE SQUARE

Sometimes a quadratic equation cannot readily be solved by factoring. However, another method called completing the square may solve it by making the unknown terms part of a square. The following equation illustrates the completing square method.

The equation is: $12P^2 = 29P - 14$

Step 1 Divide both sides of the equation by the coefficient of the unknown term that is squared. Do not simplify by actually dividing!

$$\frac{(12P^2)}{12} = \frac{29P}{12} - \frac{14}{12}$$

Step 2 Group the unknown terms together on the left side of the equal sign and place the constant term on the right. Simplify only the term that is squared.

$$P^2 - \frac{29P}{12} = -\frac{14}{12}$$

Step 3 Select a value equal to one-half the coefficient used for the second term, square this value and add it to both sides of the equation. In our example we keep the numerator (29) and double the denominator (12) before squaring and adding.

$$P^2 - \frac{29P}{12} + \left(\frac{29}{24}\right)^2 = \left(\frac{29}{24}\right)^2 - \frac{14}{12}$$

Step 4 Find the factors of the quadratic expression written on the left of the equal sign.

$$\left(\frac{P-29}{24}\right)\left(\frac{P-29}{24}\right) = \left(\frac{29}{24}\right)^2 - \frac{14}{12}$$

Step 5 Find the common denominator for the terms on the right side of the equal sign.

$$\left(\frac{P-29}{24}\right)\left(\frac{P-29}{24}\right) = \left(\frac{29}{24}\right)^2 - \frac{24(2)}{24(2)}\frac{14}{12}$$

$$\left(\frac{P-29}{24}\right)\left(\frac{P-29}{24}\right) = \left(\frac{29}{24}\right)^2 - \frac{672}{(24)^2}$$

$$\left(\frac{P-29}{24}\right)\left(\frac{P-29}{24}\right) = \frac{841\text{-}672}{(24)^2}$$

Step 6 Take the square root of each term in the equation.

$$\sqrt{\left(\frac{P-29}{24}\right)\left(\frac{P-29}{24}\right)} = \sqrt{\frac{841\text{-}672}{(24)^2}}$$

$$\frac{P-29}{24} = \sqrt{\frac{841\text{-}672}{(24)}}$$

$$\frac{P-29}{24} = \pm\frac{13}{24}$$

Step 7 Solve for the two unknowns by transposing and performing the indicated mathematical operations.

$$P = \frac{29}{24} + \frac{13}{24} = +\frac{42}{24} = \frac{7}{4} \textit{answer \#1}$$

$$P = \frac{29}{24} - \frac{13}{24} = +\frac{16}{24} = \frac{2}{3} \textit{answer \#2}$$

2.2-10D   SOLVING QUADRATIC EQUATIONS BY A QUADRATIC FORMULA

A third popular method used to solve quadratic equations is one that employs the quadratic formula which solves for the unknown without too much effort. The formula is:

$$X = -B \pm\sqrt{\frac{B^2 - 4AC}{2A}}$$

Since quadratic equations resemble the familiar $AX^2 + BX + C$ form, application of this formula is quite simple after the equation is rearranged because the values for A, B, and C are substituted into it. To illustrate this we will solve the following equation:

$$E = \frac{1}{E+1} + 3$$

First we must arrange the equation in the quadratic form $AX^2 + BX + C = 0$ form.

$$FROM \quad E = \frac{1}{E+1} + 3$$

$$E - 3 = \frac{1}{E+1}$$

$$(E - 3)(E + 1) = 1$$

$$E^2 - 2E - 3 = 1$$

$INTO \quad E^2 - 2E - 4 = 0 \; (quadratic \; form \; AX^2 + BX + C = 0)$

Second, we write the quadratic formula and then substitute the coefficients for A, B, and C.

$$X = -B \pm \frac{\sqrt{B^2 - 4AC}}{2A}$$

$$OR \quad X = \frac{-(-2) \pm \sqrt{(-2)^2 - 4(1)(-4)}}{2(1)}$$

$$X = \frac{2 \pm \sqrt{4 + 16}}{2} = \frac{2 \pm \sqrt{20}}{2} = \frac{2 \pm \sqrt{(4)(5)}}{2}$$

$$X = \frac{2 \pm 2\sqrt{5}}{2} \quad or \; X = 1 \pm \sqrt{5}$$

2.2-10E   SOLVING HIGHER ORDER EQUATIONS HAVING QUADRATIC FORM

Higher order equations of the $3R^4 - R^2 - 25 = 0$ variety can be solved using the quadratic formula if the original equation is simplified. If we substituted another unknown say (W) for R^2, then R equals $\sqrt{W}$ and the following would be true.

instead of: $3R^4 - R^2 - 25 = 0$

we have: $3W^2 - W - 25 = 0$ ($AX^2 + BX + C$ form)

If we write the quadratic formula and substitute the values we solve for the new unknown and then for the original unknown.

$$X = \frac{-B \pm \sqrt{B^2 - 4AC}}{2A} \quad or \; W = \frac{-1 \pm \sqrt{(-1)^2 - 4(3)(-25)}}{2(3)}$$

$$W = \frac{-1 \pm \sqrt{1 + 300}}{6}, \; Then: W = \frac{-1 + 17.35}{6} = \frac{16.35}{6} = 2.73$$

$$And: W = \frac{-1 - 17.35}{6} = \frac{-18.35}{6} = -3.06$$

Since earlier we said $R = \sqrt{W}$ then the values for the unknown letter R is:

$$R = \sqrt{2.73} \quad and \quad R = \sqrt{-3.06}$$

2.2-11   Simultaneous Equation Solutions

Often in electronics, two or more equations must be used when describing circuitry operation. Since an equation having only one unknown cannot be written and still adequately describe the particular phenomenon, simultaneous equations are used. Two or more equations satisfied by the same values of the unknown will be illustrated in three separate ways. The first will be using addition and subtraction methods, the second is solved by substitution, and the third method is that of comparison.

2.2-11A   SOLVING SIMULTANEOUS EQUATIONS BY ADDITION OR SUBTRACTION

Simultaneous equations may be added or subtracted thereby eliminating all but one unknown when following the following procedures. The equations used will describe the voltages in a two-source network. The unknown will be the two separate currents flowing in their circuit. The circuit we will describe, using two equations, will be:

Figure 2-4

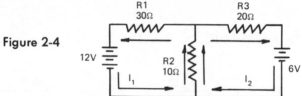

Step 1: Write the equations for the voltages around the two closed loops created when the currents flow through the resistors. Keep the variable quantities on the left side and the constant terms on the right of the equal sign.

The first equation:	*The second equation*
$E_{R_1} + E_{R_2} = 12V$	$E_{R_2} + E_{R_3} = 6V$
or: $30I_1 + 10I_1 + 10I_2 = 12$	$10I_1 + 10I_2 + 20I_2 = 6$
simplified: $40I_1 + 10I_2 = 12$	$10I_1 + 30I_2 = 6$

Step 2: Arrange the two equations as if adding or subtracting the two.

$$40I_1 + 10I_2 = 12$$
$$\underline{10I_1 + 30I_2 = \ \ 6}$$

Step 3: Multiply either equation by a number which will allow one of the like terms to cancel when the equations are subtracted. In our case we will multiply the second equation by 4 and will subtract this equation from the other. Rearrange, if necessary, to keep the unknown positive.

$$40I_1 + 120I_2 = 24 \text{ (second equation)}$$
$$\underline{-(40I_1 + \ \ 10I_2 = 12) \ \text{ (first equation)}}$$
$$+ \ 110I_2 = 12$$

Step 4: Solve for the unknown.

$$110I_2 = 12$$
$$I_2 = \frac{12}{110} = 0.109 \text{ amperes}$$

Step 5: Plug in this unknown into one of the original equations to solve for the other current.

(First equation)

$$40I_1 + 10I_2 = 12$$
$$40I_1 + 10(0.109) = 12$$
$$40I_1 + 1.09 = 12$$
$$40I_1 = 10.91$$
$$I_1 = \frac{10.91}{40} = 0.2727 \text{ amperes}$$

Step 6: Check the solutions by substituting the two current values into one or both of the original equations.

$$\textit{Equation Two} \qquad 10I_1 + 30I_2 = 6$$
$$10(0.2727) + 30(0.109) = 6$$
$$2.727 + 3.270 = 6$$
$$5.997 \approx 6$$

Naturally in this case we did not carry out our calculations far enough to equal the 6 volts, but for practice sake, it is close enough.

2.2-11B   SOLVING SIMULTANEOUS EQUATIONS BY SUBSTITUTION

Two simultaneous equations having the same unknown quantity or quantities may be solved by substituting the value of one unknown in one equation for the same unknown in the other equation. The previous example for a two-source network will be used solving for the currents.

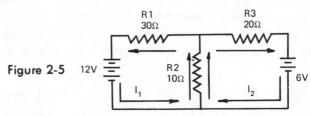

Figure 2-5

Step 1: Write the equations describing the voltages around the closed loops shown in Figure 2.5.

First equation   *Second equation*
$$40I_1 + 10I_2 = 12 \qquad\qquad 10I_1 + 30I_2 = 6$$

Step 2: Solve one equation for one unknown in terms of the other unknown.

$$\textit{First equation} \quad 40I_1 + 10I_2 = 12$$
$$40I_1 = 12 - 10I_2$$
$$I_1 = \frac{12 - 10I_2}{40} = \frac{6 - 5I_2}{20}$$

Step 3: Substitute this value of the unknown for the same unknown in the second equation and solve for one unknown.

Second equation $10I_1 + 30I_2 = 6$

$$10\left(\frac{6 - 5I_2}{20}\right) + 30I_2 = 6$$

$$60 - 50I_2 + 600I_2 = 120$$

$$60 + 550I_2 = 120$$

$$550I_2 = 60$$

$$I_2 = 0.109 \text{ amperes (one answer)}$$

Step 4: Substitute the value for the unknown just solved for into either one of the two original equations, and solve for the other unknown.

Second equation $10I_1 + 30I_2 = 6$

$$10I_1 + 30(0.109) = 6$$

$$10I_1 + 3.270 = 6$$

$$10I_1 = 2.73$$

$$I_1 = 0.273 \text{ amperes (answer)}$$

Step 5: Check the answers by substituting the two unknown values into one or both of the original equations.

Second equation check $10I_1 + 30I_2 = 6$

$$10(0.273) + 30(0.109) = 6$$

$$2.73 + 3.27 = 6$$

$$6 = 6$$

2.2-11C   SIMULTANEOUS EQUATION SOLUTIONS BY COMPARISON

An unknown can be eliminated in a simultaneous equation when both equations are solved for the same unknown quantity. Since these unknowns must satisfy both equations, they are said to be equal. It is upon this basis that we will solve the same equations used in the two previous examples, showing how to solve them by comparison methods.

Step 1: Write the equations for the circuitry illustrated in Figure 2.5.

First equation	*Second equation*
$40I_1 + 10I_2 = 12$	$10I_1 + 30I_2 = 6$

Step 2: Pick one common unknown in both equations and solve for it in terms of the other unknown.

First equation

$$40I_1 + 10I_2 = 12$$

$$40I_1 = 12 - 10I_2$$

$$I_1 = \frac{12 - 10I_2}{40}$$

$$I_1 = \frac{6 - 5I_2}{20}$$

Second equation

$$10I_1 + 30I_2 = 6$$

$$10I_1 = 6 - 30I_2$$

$$I_1 = \frac{6 - 30I_2}{10}$$

$$I_1 = \frac{3 - 15I_2}{5}$$

Step 3: Since both unknowns are equal to each other we can solve for the other unknown in the following manner.

First equation I_1 = Second equation I_1

OR: $\dfrac{6-5I_2}{20} = \dfrac{3-15I_2}{5}$

$30-25I_2 = 60-300I_2$

$30-25I_2 + 300I_2 = 60$

$275I_2 = 30$

$I_2 = \dfrac{30}{275}$

$I_2 = 0.109$ ampere (first answer)

Step 4: Substitute this quantity for the same unknown in either of the original equations and then solve for the other unknown quantity.

First equation $40I_1 + 10I_2 = 12$
$40I_1 + 10(0.109) = 12$
$40I_1 = 12-1.09$
$I_1 = \dfrac{10.91}{40}$
$I_1 = 0.273$ amperes

Step 5: To verify the solutions, solve either or both equations using the values just calculated.

Equation two $10(I_1) + 30(I_2) = 6$
$10(0.273) + 30(0.109) = 6$
$2.73 + 3.27 = 6$
$6 = 6$ check

Simultaneous equations having three or more unknowns in three or more equations can be solved using the general methods outlined in 2.2-11A through 2.2-11C.

A common step sequence for solving three unknown quantities in three equations is as follows:

Step 1: Use only two of the three equations to form an equation having only two known quantities.

Step 2: Use the equation written for Step 1 and one of the original equations to form a second equation having only two unknown quantities. The second equation should have the same unknowns as the first equation having two unknown quantities.

Step 3: Take the two simplified equations with two unknown quantities and solve for one unknown in terms of the other unknown.

Step 4: Substitute the results obtained in Step 3 for the unknown quantities in any of the original equations solving for the third unknown.

Step 5: Check your answers by substituting your values into one or all of the three original equations.

2.3   PLANE GEOMETRY

This portion of Chapter 2 will provide you with a condensed view of Geometry dealing with the rules. terminology, postulates and facts needed to solve electronic problems with mathematical wisdom.

2.3-1   Postulates

Sometimes things may be accepted without proof. In geometry, principles just accepted are called postulates and serve to develop rules for arithmetic and algebraic relationships. The following postulates are those most often relied upon for all mathematical processes.

1. The shortest distance between two points in a straight line.
2. Only one straight line can be drawn between any two points.
3. Two straight lines will not intersect at more than one point.
4. You may extend a straight line to any distance.
5. A geometric figure can be moved from one position to another without altering its basic form or magnitude.
6. All 90 degree angles (right angles) are equal.
7. All straight lines have a straight angle of 180 degrees.
8. In a flat surface (plane) only one perpendicular to a line can be drawn from a point placed on either side of the plane.
9. Angles having the same supplement are equal.
10. Angles having the same complement are equal.
11. Vertical angles are equal.
12. A line segment can be bisected at only one point.
13. All angles can be bisected by only one line.
14. All the sides of a square are equal.
15. All radii of the same circle are equal.
16. All diameters of the same circle are equal.
17. A straight line can intersect a circle at only two points.
18. If the two points where a straight line intersects a circle coincide, the line is tangent to the circle.
19. A circle intersects another circle only at two points.
20. The diameter of a circle bisects the circle.
21. There are 360 degrees in a square and in a circle.

2.3-2   Angles

Angles are generally divided into five separate groups, as illustrated in Figure 2-6. The first is called the *right angle* since a line drawn perpendicular to another line forms a 90 degree angle. The second group is called an *acute angle* because the angle formed is less than 90 degrees. The third group, called an *obtuse angle*, has an angle that is greater than 90 degrees, but less than 180 degrees or that angle for a straight line.

The fourth group is the *straight line*. Its angle is formed when the sides extend in opposite directions from the *vertex* or point where the lines meet. The fifth group, called a *reflex angle,* has more than 180 degrees, but less than 360 degrees for its angle.

Since two straight lines that meet at a point can and do form angles, the sides and the angle formed can be used for angle identification. Angles formed by two straight lines are normally represented by the letter symbol "O", the Greek letter theta (Θ), the geometry angle symbol ($\angle$) plus the letter O or Θ, or by writing the assigned letters representing the line terminates preceeded by the angle symbol ($\angle$AOC). These concepts are shown in Figure 2.6 for the five angle groups.

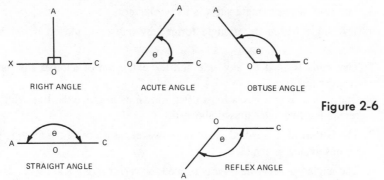

Figure 2-6

Since all of the angles illustrated in Figure 2.6 share a common vertex and common sides, the angles are called *adjacent angles*. The right angle example shows angles "XOA" and "AOC" having a small square at the point where line "AO" is drawn. These squares represent an angle of 90 degrees or that the line is perpendicular to line "XC."

The application of geometry involves three basic angle relationships termed *complementary angles, supplementary angles,* and *vertical angles.*

Complementary angles are formed whenever two angles added together equal a right angle of 90 degrees. Angle AOC shown in Figure 2.7 equals 90 degrees, and angles AOB plus BOC when combined also equal the required 90 degrees. Hence, the complementary angles are AOB and BOC.

The supplementary angle in Figure 2.7 is formed when the two angles added together equal a straight angle having 180 degrees. Angle DOF equals 180 degrees and angle "DOE" plus "EOF" when added together also equal 180 degrees. The supplementary angles are "DOE" and "EOF."

Vertical angles consist of two lines that intersect in the same manner as illustrated in Figure 2.7. The opposite angles formed are called vertical angles and are recognized as being "IOH," "GOJ," and angles "IOG" and "HOJ."

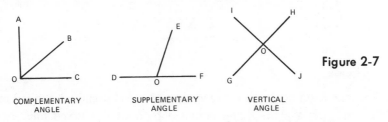

Figure 2-7

2.3-3   Triangles

The terms and definitions for all triangles are listed here in brief form, but will serve your needs in geometry, trigonometry, and other related maths for electronic applications.

1. The portion of a plane surrounded by three straight lines is called a *triangle*.
2. The straight lines forming the triangle are called its *sides*.
3. The sum of the triangle sides is its *perimeter*.
4. The angle within the triangle formed by any two sides is its *interior angle*.
5. The angle formed by any side and the extension of another side is an *exterior angle*.
6. One side of a triangle is most often drawn as a horizontal line with the remaining two sides drawn above it.
7. The bottom side is called the *base*; however, any side of the triangle can be designated as the base.
8. The angle opposite the base is termed the *vertex angle* and serves as the vertex of the triangle.
9. The perpendicular distance from the vertex of the triangle to its base is called the *altitude*.
10. If a triangle is a right triangle, the side opposite the right angle (the longest side) is termed the *hypotenuse*.

The five general types of triangles illustrated in Figure 2-8 are briefly described in the following way. The *acute triangle* has its interior angles somewhat less than a right angle which has 90 degrees. The *right triangle* has one right angle and two acute angles. The *obtuse triangle* has one obtuse angle (greater than 90 degrees) and two acute angles. The *isosceles triangle* can be identified easily because it has two equal sides and two equal angles. The last triangle, an *equilateral triangle* or equiangular triangle, has three equal sides and three equal angles.

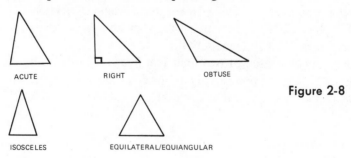

Figure 2-8

2.3-4   Circles

The simple circle you and I recognize is geometrically described in the following similar manner: a plane figure bounded by a single curved line, every point of which is equally distant from the point at the center of the figure. Now that you have the idea of what a circle really is, we will provide the terminology in a simpler form. Figure 2-9 illustrates the following facts.

1. The curved line forming the circle is called the *circumference*.

2. A straight line drawn between the center of the circle and any point on its circumference is termed the *radius*.

3. A chord is a straight line (called straight line segment) which joins any two points of the circle's circumference.

4. A chord which happens to pass through the center of the circle is called the *diameter*.

5. The diameter is actually twice the length of the circle's radius.

6. A straight line which touches a circle, but does not cross its circumference, is a *tangent* of the circle.

7. The point at which the tangent touches the circumference of the circle is called the *point of tangency*.

8. A straight line which cuts a circumference in two points but not at a circle's center is a *secant* of that circle.

9. A *central angle* within a circle is an angle which has its vertex at the center of the circle and whose sides are the radii of the circle.

10. An *inscribed angle* within a circle is an angle whose vertex is in the circumference of the circle and whose sides are chords.

11. Circles having a common center are called *concentric circles*.

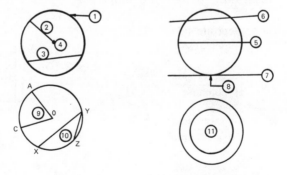

Figure 2-9

1. Circumference	7. Tangent
2. Radius	8. Point of tangency
3. Chord	9. Central angle (AOC)
4. Center	10. Inscribed angle (XYZ)
5. Diameter	11. Concentric circles
6. Secant	

2.4   TRIGONOMETRY

This portion of Chapter 2 deals with the specialized mathematics termed trigonometry. Trigonometry applies the rules and laws of arithmetic for its numbers, the rules and laws of algebra in its equations, the rules and laws of geometry and the rules and laws of trigonometric relationships between triangle angles and their sides during application. Since most of the important rules, facts, and laws serving as the foundation for this specialized math have been uncovered earlier in this chapter, the following rules, facts, and guidelines will be introduced in a concise manner having a minimum amount of previous data review.

2.4-1   Trigonometric Relations

A high percentage of electronic calculations involving alternating current must be expressed mathematically using right triangle facts and angle designations to accurately describe the electronic phenomenon. Relationships used to describe these phenomena are based upon the acute angle (the angle equal to or less than 90 degrees) and the sides of a right triangle. There are six functions of the right triangle for angles employed in trigonometry. The names for the six functions and their abbreviations are:

1. Sine (sin)	4. Cotangent (cot)
2. Cosine (cos)	5. Secant (sec)
3. Tangent (tan)	6. Cosecant (csc)

The functions of the right triangle provide a usable number, after some calculation, which corresponds to an angle measurement in degrees. The illustration in Figure 2-10 will serve as an aid, providing meaning to the function name and angle references.

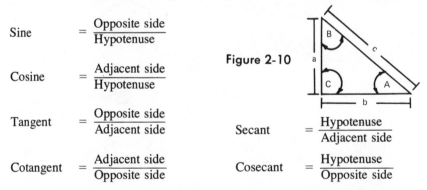

$$\text{Sine} = \frac{\text{Opposite side}}{\text{Hypotenuse}}$$

$$\text{Cosine} = \frac{\text{Adjacent side}}{\text{Hypotenuse}}$$

Figure 2-10

$$\text{Tangent} = \frac{\text{Opposite side}}{\text{Adjacent side}}$$

$$\text{Secant} = \frac{\text{Hypotenuse}}{\text{Adjacent side}}$$

$$\text{Cotangent} = \frac{\text{Adjacent side}}{\text{Opposite side}}$$

$$\text{Cosecant} = \frac{\text{Hypotenuse}}{\text{Opposite side}}$$

The longer side marked with the smaller letter "c" illustrated in Figure 2-10 represents the hypotenuse. The hypotenuse, as you remember, is always the longest side of any right triangle. The terms "opposite" or "adjacent" sides used in describing the trigonometric functions have movable meanings depending upon which of the three angles in the triangle you are talking about. For example, the sine function for angle

"A" has as its opposite side the letter "a" while the sine function for angle "B" has the letter "b" representing its opposite side and in like manner the sine function for angle "C" would have the letter "c" indicating its opposite side.

2.4-2   Trigonometric Application

Application of any one function would not be possible unless someone, someplace, at sometime developed its relationships. Rather than sharing with you the tedious calculations essential to generate such truths, the following formulas can be applied directly to trigonometry-related problems if you know the values for the indicated side and angle for your problem. Use Figure 2-10 as a reference to aid in recognizing the side or angle.

Values you know (sides and angles)	Formulas to use in solving for the unknown		
1. side a, side b	$c = \sqrt{a^2 + b^2}$	$\mathrm{Tan\ A} = \dfrac{a}{b}$	$\angle\,B = 90° - \angle\,A$
2. side a, hypotenuse c	$b = \sqrt{c^2 - a^2}$	$\mathrm{Sin\ A} = \dfrac{a}{c}$	$\angle\,B = 90° - \angle\,A$
3. side b, hypotenuse c	$a = \sqrt{c^2 - b^2}$	$\mathrm{Sin\ B} = \dfrac{b}{c}$	$\angle\,A = 90° - \angle\,B$
4. Hypotenuse c, angle B	$b = c\ \mathrm{Sin\ B}$	$a = c\ \mathrm{Cos\ B}$	$\angle\,A = 90° - \angle\,B$
5. Hypotenuse c, angle A	$b = c\ \mathrm{Cos\ A}$	$a = c\ \mathrm{Sin\ A}$	$\angle\,B = 90° - \angle\,A$
6. side b, angle B	$c = \dfrac{b}{\mathrm{Sin\ B}}$	$a = b\ \mathrm{Cot\ B}$	$\angle\,A = 90° - \angle\,B$
7. side b, angle A	$c = \dfrac{b}{\mathrm{Cos\ A}}$	$a = b\ \mathrm{Tan\ A}$	$\angle\,B = 90° - \angle\,A$
8. side a, angle B	$c = \dfrac{a}{\mathrm{Cos\ B}}$	$b = a\ \mathrm{Tan\ B}$	$\angle\,A = 90° - \angle\,B$
9. side a, angle A	$c = \dfrac{a}{\mathrm{Sin\ A}}$	$b = a\ \mathrm{Cot\ A}$	$\angle\,B = 90° - \angle\,A$

It should be pointed out that this information is *only* valid when applied to right triangle problems and that if you know any two measurements (angle and side, or two sides) all of the other data pertinent to that right triangle problem can be solved. To illustrate the usefulness of the listed formulas, we will show how to solve for the three unknowns when side "a" and side "b" are given.

The electronic problem involves a 400 ohm resistor which is connected in series with an inductor having an inductive reactance of 300 ohms. Naturally, the question would be what is the total impedance of this circuit, but we won't stop there because we can figure out what the other two angles would be also, as you will soon see.

The best way to begin any reactive AC problem is first to draw the circuit, second draw the right triangle equivalent for the circuit, and then solve the problem using appropriate rules and laws of mathematics. The circuit and impedance triangle, shown in Figure 2-11, illustrates the value of resistance and inductive reactance for the right triangle sides. The right triangle used for Figure 2-10 is also shown in Figure 2-11 to illustrate similarities of previously defined materials.

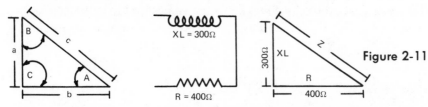

Figure 2-11

Since side "a" represents 300 ohms, side "b" represents 400 ohms, and side "c" the total impedance, direct application can be made as this mathematical process indicates.

1. Since $c = \sqrt{a^2 + b^2}$ then: $Z = \sqrt{(XL)^2 + (R)^2}$

$$Z = \sqrt{(300)^2 + (400)^2}$$
$$Z = \sqrt{90,000 + 160,000}$$
$$Z = \sqrt{250,000}$$
$$Z = 500 \text{ ohms}$$

The total impedance for the series circuit is 500 ohms. The data for the problem was plugged into the formulas in row #one of the previous data chart since we knew the value for both sides of the triangle. Angle "A" or angle "B" could be determined in a similar manner once the data for your particular problem is substituted into the appropriate formula. For example, we shall show how to apply the data found in Figure 2-11 when solving for angles "A" and "B."

2. Since $\text{Tan "A"} = \dfrac{a}{b}$ (refer to row #1, 3rd column of formulas)

Then: $\text{Tan "A"} = \dfrac{300}{400}$ (refer to Figure 2-11)

$\text{Tan "A"} = 0.75$

This number, when looked up in a Natural Trigonometric Functions Table, will provide us with the exact angle value or its approximate angle for the tangent function equaling 0.75. If you were to find this number in the Natural Trigonometric Function Chart, only an approximate angle between 36 degrees-50 minutes and 37 degrees could be found. The more exact answer often demanded for electronic calculation must be determined by interpolation, a process that will be illustrated after we demonstrate how to solve for the "B" angle.

Angle "B" can be found by applying the facts of the problem to the formula listed in row #1, 4th column or by taking the tangent function of it. Either method should result in the same angle value as indicated next.

3. $\angle$ "B" = 90° minus $\angle$ A   (OR) Tan "B" $= \dfrac{b}{a}$

$\angle$ "B" = 90° minus about 37°   Tan "B" $= \dfrac{400}{300}$

$\angle$ "B" = approximately 53°   Tan "B" = 1.333

Angle "B" equals an angle of somewhere between 36 degrees 50 minutes and 37 degrees. Since this is the same approximate value obtained for angle "A," the mathematical process called interpolation need only be done once because it will provide us with the same answer for both angles as you shall see.

It should be pointed out that trigonometric tables needed for trigonometric values and interpolation often list angles ranging from 0 to 45 degrees on the left column reading them from top to bottom. Angles ranging from 45 to 90 degrees are listed in the right column and are read from bottom to top. The Natural Trigonometric Functions Table for the numbers we are interested in resembles the following table data.

0 to 45 DEGREES	SIN	COS	TAN	COT	SEC	CSC	
36 - 00'	0.5878	0.8090	0.7265	1.376	1.236	1.701	54 - 00'
36 - 50'	0.5995	0.8004	0.7490	1.335	1.249	1.668	53 - 10'
37 - 00'	0.6018	0.7986	0.7536	1.327	1.252	1.662	53 - 00'
	COS	SIN	COT	TAN	CSC	SEC	45 to 90 DEGREES

The corresponding angles for the number 0.75 representing the Tan "A" function and the number 1.333 representing the Tan "B" function is determined in this way. Tangent "A" equaling 0.75 will be interpolated first.

1. Find the two closest numbers for the tangent function and their angles. Since the two closest number 0.7490 and 0.7536 are listed under the tangent function listed at the top of the chart, the angles are those shown on the left of the table.

2. Record this data in the following way.

36° 50'   = 0.7490

Tan "A"   = 0.7500

37° 00'   = 0.7536

3. Then, if 36° 50' is less than the unknown angle, and 37° 00' is larger than the unknown, the solution can be found by knowing that 37 degrees also equals 36 degrees, 59 minutes, 60 seconds or 36° 59' 60". We will change this to another equivalent of 36° 58' 120" for greater accuracy.

$$\frac{37° \ 00' \ - \ \text{Tan "A"}}{37° \ 00' \ - \ 36° \ 50'} = \frac{0.7536 - 0.7500}{0.7536 - 0.7490}$$

$$\frac{36° \ 59' \ 60'' \ - \ \text{Tan "A"}}{36° \ 60' \ - \ 36° \ 50'} = \frac{0.0036}{0.0046}$$

$$\frac{36° \ 59' \ 60'' \ - \ \text{Tan "A"}}{10'} = 0.78$$

Tan "A" = 36° 58' 120" − (10)(0.78)

Tan "A" = 36° 58' 120" − 7' 80"

Tan "A" = 36° 51' 40"

Angle "A" illustrated in Figure 2-11 is 36 degrees, 51 minutes, 40 seconds, which is somewhat closer than our estimate of somewhere between 36 degrees, 50 minutes and 37 degrees. Angle "B" can be evaluated in the same general manner as angle "A" or for simplicity, since we know angle "C" is 90 degrees and angle "A" is 36° 51' 40" then:

Angle "B" = 90° − angle "A"

∠ "B" = 89° 59' 60" − 36° 51' 40"

∠ B = 53° 08' 20"

The right triangle problem, illustrated in Figure 2-11, can be checked for angle accuracy since in any triangle the sum of its angles will equal 180 degrees. If you add the three angles, "A", "B" and "C" together, the result is 179° 59' 60" or the equivalent of 180 degrees.

2.4-3   Quadrant Signs and Function Relations

The circle, illustrated in Figure 2-12, will serve as a basis for quadrant division and trigonometric function angle relationships. The circle is divided in half twice by two axes. The axis drawn north and south is called the "Y" axis while the axis drawn east and west is called the "X" axis. These two axes divide the circle into four equal quadrants consisting of 90 degrees each. Quadrant I has angles ranging from 0 to 90 degrees, quadrant II has angles from 90 to 180 degrees, quadrant III from 180 to 270 degrees, and the fourth quadrant from 270 to 360 degrees.

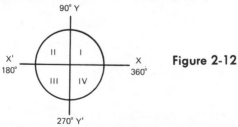

Figure 2-12

The signs for the X and Y magnitudes would be positive in the first quadrant. The signs for the second quadrant would be mixed; Y is positive and X' is negative. Both Y' and X' in the third quadrant would be negative. The fourth quadrant magnitudes for X and Y′ would be mixed, Y′ is negative while X is positive. Rather than remembering the above facts and relating them to different functions and trying to recall the range of numbers corresponding to the angles for each quadrant, the following chart provides instant recall with less mental fatigue. Figure 2-12 should be referred to for quadrant and angle limits.

Trigonometric Function	Quadrant I 0 to 90°	Quadrant II 90 to 180°	Quadrant III 180 to 270°	Quadrant IV 270 to 360°
Sine	Positive 0 to 1	Positive 1 to 0	Negative 0 to −1	Negative −1 to 0
Cosine	Positive 1 to 0	Negative 0 to −1	Negative −1 to 0	Positive 0 to 1
Tangent	Positive 0 to ∞	Negative ∞ to 0	Positive 0 to ∞	Negative ∞ to 0
Cotangent	Positive ∞ to 0	Negative 0 to ∞	Positive ∞ to 0	Negative 0 to ∞
Secant	Positive 1 to ∞	Negative ∞ to −1	Negative −1 to ∞	Positive ∞ to −1
Cosecant	Positive ∞ to 1	Positive 1 to ∞	Negative ∞ to −1	Negative −1 to ∞

Since trigonometric function tables use only those angles found within quadrant I, we must relate angles greater than 90 degrees to quadrant one facts in order to use Trigonometric Tables. With application of the data enclosed in the previous chart, we shall record what each function means in three quadrants. The first quadrant is omitted because all of its functions are positive and can be found directly in the Trigonometric Function Tables.

The trigonometric functions for an angle (ϕ) in the second quadrant are the same functions as for an angle represented by $180° - \phi$. This will allow us to use the functions listed in the Trigonometric Tables. Follow the functions listed in the previous chart, paying attention to the sign given for each function. *The second quadrant functions are:*

$$\text{Sin } \phi = \text{Sin } (180° - \phi) \qquad \text{Cot} = -\text{cot } (180° - \phi)$$
$$\text{Cos } \phi = -\text{Cos } (180° - \phi) \qquad \text{Sec} = -\text{sec } (180° - \phi)$$
$$\text{Tan } \phi = -\text{Tan } (180° - \phi) \qquad \text{Csc} = \text{Csc } (180° - \phi)$$

The functions for the angle ϕ in the *third quadrant* are as follows. Note the sign similarity from the chart previously used.

$$\text{Sin } \phi = -\text{Sin } (\phi - 180°) \qquad \text{Cot } \phi = \text{Cot } (\phi - 180°)$$
$$\text{Cos } \phi = -\text{Cos } (\phi - 180°) \qquad \text{Sec } \phi = -\text{Sec } (\phi - 180°)$$
$$\text{Tan } \phi = \text{Tan } (\phi - 180°) \qquad \text{Csc } \phi = -\text{Csc } (\phi - 180°)$$

If you have been following the chart, you should be able to write the functions for the angle ϕ found in the *fourth quadrant*. They are as follows:

$$\text{Sin } \phi = -\text{Sin } (360°-\phi) \qquad \text{Cot } \phi = -\text{Cot } (360°-\phi)$$
$$\text{Cos } \phi = \text{Cos } (360°-\phi) \qquad \text{Sec } \phi = \text{Sec } (360°-\phi)$$
$$\text{Tan } \phi = -\text{Tan } (360°-\phi) \qquad \text{Csc } \phi = -\text{Csc } (360°-\phi)$$

2.4-4   Functions of Angles Greater Than 360 Degrees

The secret of solving problems involving angles greater than 360 degrees is based upon the fact that the trigonometric functions are the same once an equivalent angle relationship is obtained. In order to understand this concept, we will show how to find the cosine function for an angle of 950 degrees.

To begin, we divide the 950 degree angle by 360 degrees so that a relationship to an angle less than 360 degrees may be obtained.

1. Divide 950° by 360°:

$$
\begin{array}{r}
2 \\
360\overline{)950} \\
720 \\
\hline
230
\end{array}
$$

The remainder of 230 degrees is the secret number since it will provide us with the same trigonometric function for 950 degrees. After observing which quadrant the remainder of 230 degrees is in, the cosine function equation can be written and solved. Since it is in the third quadrant, the sign for the function is negative and is written as follows:

2. $\text{Cos } 230° = -\text{Cos } (230°-180°)$

$\text{Cos } 230° = -\text{Cos } 50°$

$\text{Cos } 230° = -0.6428$

2.4-5   Combined Angle Function Formulas

You will, from time to time, be required to solve electronic problems having trigonometric functions of more than one angle. These angles might be added, subtracted, divided, or multiplied. Since direct mathematical processes cannot be applied, another method put in formula form will aid you when the conditions arise. The following formulas are listed according to the mathematical process needed for adding or subtracting angles, etc.

2.4-5A   SUM OF TWO ANGLES

$$\text{Sin } (A + B) = \text{Sin } A \text{ Cos } B + \text{Cos } A \text{ Sin } B$$
$$\text{Cos } (A + B) = \text{Cos } A \text{ Cos } B - \text{Sin } A \text{ Sin } B$$
$$\text{Tan } (A + B) = \frac{\text{Tan } A + \text{Tan } B}{1 - \text{Tan } A \text{ Tan } B}$$
$$\text{Cot } (A + B) = \frac{\text{Cot } A \text{ Cot } B - 1}{\text{Cot } B + \text{Cot } A}$$

2.4-5B   DIFFERENCE OF TWO ANGLES

$$\text{Sin } (A-B) = \text{Sin } A \text{ Cos } B - \text{Cos } A \text{ Sin } B$$

$$\text{Cos } (A-B) = \text{Cos } A \text{ Cos } B + \text{Sin } A \text{ Sin } B$$

$$\text{Tan } (A-B) = \frac{\text{Tan } A - \text{Tan } B}{1 + \text{Tan } A \text{ Tan } B}$$

$$\text{Cot } (A-B) = \frac{\text{Cot } A \text{ Cot } B + 1}{\text{Cot } B - \text{Cot } A}$$

2.4-5C   TWICE THE ANGLE

The following formulas were generated by letting A equal B, and instead of writing the function (sine, cosine, etc.) and then $(A + A)$ its equivalent is written:

$$\text{Sin } (A + A) = \text{Sin } 2A$$
$$\text{Sin } 2A = 2 \text{ Sin } A \text{ Cos } A$$

$$\text{Cos } (A + A) = \text{Cos } 2A$$
$$\text{Cos } 2A = \text{Cos }^2A - \text{Sin }^2A$$
$$\text{Cos } 2A = 1 - 2 \text{ Sin }^2A$$
$$\text{Cos } 2A = 2 \text{ Cos }^2A - 1$$

$$\text{Tan } (A + A) = \text{Tan } 2A$$
$$\text{Tan } 2A = \frac{2 \text{ Tan } A}{1 - \text{Tan}^2A}$$
$$\text{Tan } 2A = \frac{2}{\text{Cot } A - \text{Tan } A}$$

$$\text{Cot } (A + A) = \text{Cot } 2A$$
$$\text{Cot } 2A = \frac{\text{Cot }^2A - 1}{2 \text{ Cot } A}$$
$$\text{Cot } 2A = \frac{\text{Cot } A - \text{Tan } A}{2}$$

2.4-6   Trigonometric Identities and Formulas

$$\text{Sin } A = \sqrt{1 - \text{Cos}^2 A}$$

$$\text{Sin } A = \sqrt{\frac{\text{Tan } A}{1 + \text{Tan}^2 A}}$$

$$\text{Sin } A = \sqrt{\frac{1}{1 + \text{Cot}^2 A}}$$

$$\text{Sin } A = 2 \text{ Sin } \tfrac{1}{2}A \text{ Cos } \tfrac{1}{2}A$$

$$\text{Sin } A = \frac{2 \text{ Tan } \tfrac{1}{2}A}{1 + \text{Tan}^2 \tfrac{1}{2}A}$$

$$\text{Sin }^2A - \text{Sin }^2B = \text{Cos }^2B - \text{Cos}^2A$$
$$\text{Sin }^2A - \text{Sin }^2B = \text{Sin } (A + B) \text{ Sin } (A - B)$$
$$\text{Sin } A \text{ Sin } B = \tfrac{1}{2} \text{ Cos } (A-B) - \tfrac{1}{2} \text{ Cos } (A + B)$$
$$\text{Sin } A \text{ Cos } B = \tfrac{1}{2} \text{ Sin } (A + B) + \tfrac{1}{2} \text{ Sin } (A - B)$$

$$2 \operatorname{Sin}^2 A = 1 - \operatorname{Cos} 2A$$

$$\operatorname{Cos} A = \sqrt{1 - \operatorname{Sin}^2 A}$$

$$\operatorname{Cos} A = \frac{1}{\sqrt{1 + \operatorname{Tan}^2 A}}$$

$$\operatorname{Cos} A = \frac{\operatorname{Cot} A}{\sqrt{1 + \operatorname{Cot}^2 A}}$$

$$\operatorname{Cos} A = \frac{1 - \operatorname{Tan}^2 \frac{1}{2} A}{1 + \operatorname{Tan}^2 \frac{1}{2} A}$$

$$\operatorname{Cos} A \operatorname{Cos} B = \frac{1}{2} \operatorname{Cos} (A - B) + \frac{1}{2} \operatorname{Cos} (A + B)$$

$$\operatorname{Cos}^2 A - \operatorname{Sin}^2 B = \operatorname{Cos}^2 B - \operatorname{Sin}^2 A$$

$$\operatorname{Cos}^2 A - \operatorname{Sin}^2 B = \operatorname{Cos} (A + B) \operatorname{Cos} (A - B)$$

$$2 \operatorname{Cos}^2 A = 1 + \operatorname{Cos}^2 A$$

$$\operatorname{Tan} A = \frac{\operatorname{Sin} A}{\operatorname{Cos} A}$$

$$\operatorname{Tan} A = \frac{1}{\operatorname{Cot} A}$$

$$\operatorname{Tan} A + \operatorname{Tan} B = \frac{\operatorname{Sin} (A + B)}{\operatorname{Cos} A \operatorname{Cos} B}$$

$$\operatorname{Tan} A - \operatorname{Tan} B = \frac{\operatorname{Sin} (A - B)}{\operatorname{Cos} A \operatorname{Cos} B}$$

$$\operatorname{Tan} A \operatorname{Tan} B = \frac{\operatorname{Tan} A + \operatorname{Tan} B}{\operatorname{Cot} A + \operatorname{Cot} B}$$

$$\operatorname{Cot} A = \frac{\operatorname{Cos} A}{\operatorname{Sin} A}$$

$$\operatorname{Cot} A = \frac{1}{\operatorname{Tan} A}$$

$$\operatorname{Cot} A + \operatorname{Cot} B = \frac{\operatorname{Sin} (A + B)}{\operatorname{Sin} A \operatorname{Sin} B}$$

$$\operatorname{Cot} A - \operatorname{Cot} B = \frac{\operatorname{Sin} (A - B)}{\operatorname{Sin} A \operatorname{Sin} B}$$

$$\operatorname{Cot} A \operatorname{Cot} B = \frac{\operatorname{Cot} A + \operatorname{Cot} B}{\operatorname{Tan} A + \operatorname{Tan} B}$$

2.5　COMPLEX NUMBERS AND VECTORS

Real and imaginary numbers, when combined, form complex numbers. These complex numbers are used extensively when performing AC calculations for series, parallel, and series-parallel circuits containing RL, RC, and RLC components. This portion of Chapter 2 will provide you with the working knowledge essential for performing mathematical operations of addition, subtraction, division, and multiplication of complex numbers. This material also applies its concepts to vectorial analysis of electronic phenomenon.

2.5-1   "j" Factor/Operator

Frequently in mathematics the "j" factor or "j" operator is encountered. The "j" factor is imaginary. Since there is no *real* number which when squared equals -1, the $\sqrt{-1}$ is used. In mathematics $\sqrt{-1}$ is usually represented as "i" but in electronics this would cause confusion with current so we use "j" which equals $\sqrt{-1}$ also.

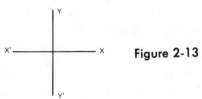

Figure 2-13

Generally positive and negative *real* numbers have graphical representation on the horizontal (x) axis while the vertical (y) axis has imaginary numbers represented, as shown in Figure 2-13. Often times the X′ is indicated with a minus (−) sign and the X with a positive (+) sign. In like manner the Y and Y′ become the + j and the − j representing the imaginary values. By using this process, both real and imaginary values can be represented on the same graph, as illustrated by Figure 2-14.

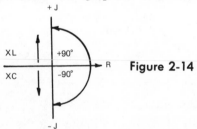

Figure 2-14

When graphically plotted, resistance is treated as a *real* number since there is no phase change. Reactances on the other hand are plotted as if they were imaginary. This method helps us to remember the phase relationship of reactances and simplifies vectorial manipulation.

2.5-2   Vectors

A vector is used to describe a quantity that has both mass and direction. The direction of the vector is indicated by an arrow, while an angle, normally called theta (Θ), provides its direction when compared to the real number axis. The mass, magnitude, or absolute value of the vector is its length. There are three general ways, methods, or forms used to provide the vectorial solutions. The three forms expressed by formulas will be outlined first; then helpful examples will follow.

The main advantage the *polar form* provides us with is an easier way to write vectorial relationships. An example of a written polar form vector is:

$$A \angle \Theta \quad \text{where:} \quad \text{A is the magnitude}$$
$$\Theta \text{ is the direction angle}$$

The *rectangular form* is commonly used when addition or subtraction is necessary, but is not generally used when division, multiplication, roots, or power processes are required. An example of the rectangular form is:

A + jB or A−jB   where: A is the real number magnitude
                           ± j is the "j" operator
                           B is the imaginary numbers' magnitude

The *trigonometric or circular form* is easily recognized because of the familiary functions included within its vector descriptions. An example of the trigonometric form is:

A (Cos Θ + j Sin Θ)   where: A is the magnitude
           OR   Θ is the direction angle
                             ± j is the "j" operator

A (Cos Θ − j Sin Θ)

2.5-3   Polar Form Vector Representation

Polar vector notation requires four basic parts which define the vector. These are: (1) magnitude, (2) angle symbol, (3) sign of the angle, and (4) number of degrees. In order to show the relationships of the four parts of a polar form described vector, we will draw a vector for an airplane flying at a speed of 500 miles per hour in a north-eastern direction. Figure 2-15 illustrates this particular vector.

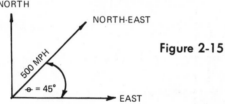

Figure 2-15

If you are vectorially alert, you will note that the magnitude is 500, the angle symbol is Θ, the sign of the angle is positive, and the number of degrees is 45. All of the four basic parts are defined and therefore this vector can be written in polar form. The complete polar notation is written as 500∠+ 45 and is read by saying 500 at positive 45 degrees.

2.5-3A   POLAR FORM MULTIPLICATION AND DIVISION PROCESSES

Since polar notation is a mathematical term, it can be treated as such during computation. If, for example, you have two vectors in polar form you can multiply the two or you can divide one by the other by following these simple rules.

Multiplication Rule
(30∠20°) (10∠20°) = 300∠40°

1. Multiply the magnitudes (30) (10) = 300
2. Add the angles 20° + 20° = 40°

Division Rule
$$\frac{30∠30°}{10∠20°} = 3∠10°$$

Rule 1. divide the magnitudes $\frac{30}{10} = 3$

Rule 2. subtract (algebraically) the angle of
the divisor from the angle of the dividend:

$$30° - 20° = 10°$$

You can add or subtract vectors in polar form *if the angles are exactly the same*. In application, the magnitude of each vector is added or subtracted and the original angle is used. Thus, if vector $10\angle + 33°$ is added to $10\angle + 33°$ the resultant is $20\angle + 33°$ or if $20\angle + 45°$ is subtracted from $50\angle + 45°$ the resultant would be $30\angle + 45°$. When vectors do not have the same exact angle, such as $6\angle + 33° + 10\angle + 45°$ or $20\angle - 90$ minus $30\angle + 90°$ the mathematical process of addition and subtraction cannot be accomplished in the above manner. You can, however, add or subtract vectors whose angles differ if you convert to the rectangular form of notation.

2.5-4   Rectangular Form Vector Representation

Vectors stated in rectangular coordinates can be readily added or subtracted since their vertical components ("j" operator) always lie on the same line while the real numbers are always on the horizontal line. The first vector example used to illustrate the polar notation will be used here also. Refer to Figure 2-16.

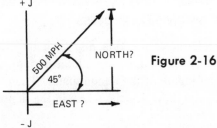

Figure 2-16

In order to describe a vector for an airplane flying at 500 mph at a 45° angle, the rectangular form becomes a trigonometric problem because we are analyzing the rate at which the airplane is traveling both north and east. To illustrate the mathematical process for the rectangular form we will solve only for the quantity marked north.

Since the side of Θ equals 45° which also equals $\frac{\text{opposite}}{\text{hypotenuse}}$

$$\text{Then: Sin } 45° = \frac{\text{north (N)}}{500}$$

$$0.707 = \frac{N}{500}$$

$$N = 500 (0.707)$$

$$N = 353.5$$

In rectangular notation we have a magnitude of 500 at a given north or "j" operator direction. The rectangular form $500 + j\,353.5$ accurately describes this data.

2.5-4A   RECTANGULAR FORM ADDITION AND SUBTRACTION PROCESS

To add or subtract vectors stated in rectangular notation, addition and subtraction are accomplished merely through algebraic manipulation. For example:

$$
\begin{array}{rcl}
6 & + & j4 \\
(+)\ \ 8 & - & j6 \\
\hline
14 & - & j2
\end{array}
\qquad \text{OR} \qquad
\begin{array}{rcl}
10 & + & j10 \\
(-)\ 14 & - & j8 \\
\hline
-4 & + & j18
\end{array}
$$

2.5-4B   RECTANGULAR FORM MULTIPLICATION PROCESS

Vectors stated in rectangular notation may be directly multiplied if we treat each vector as a binomial. For example we will multiply a vector of $5 + j4$ by $10-j4$.

$$
\begin{array}{l}
5 + j4 \\
10 - j4 \\
\hline
50 + j40 \\
\quad\ - j20 - j^2 16 \\
\hline
50 + j20 - j^2 16 \ \text{(multiplication completed)}
\end{array}
$$

Since $j^2 = -1$ then:

$$50 + j20 - (-1)(16)$$
$$50 + j20 + 16 = 66 + j20 \ \text{(simplified answer)}$$

2.5-4C   RECTANGULAR FORM DIVISION PROCESS

Vectors stated in rectangular notation may also be divided. In our example vector $6 + j5$ will be divided by vector $4 + j2$. Both the denominator and the numerator must be multiplied by the conjugate of the denominator. The conjugate is determined simply by changing the one sign between the terms in the denominator. All of this is accomplished as follows:

$$\frac{6 + j5\ (4-j2)}{4 + j2\ (4-j2)} = \frac{24 + j8 - j^2 10}{16 - j^2 4} = \frac{24 + j8 - (-1)(10)}{16 - (-1)(4)} = \frac{34 + j8}{20}$$

$1.7 + j0.4$ is the simplified answer.

2.5-5   Trignometric/Circular Form

We will re-employ the vector illustrated for the airplane whose speed was 500 mph in a north-east direction to tie together what has been discussed. Polar form, as you will recall, was simply the vector magnitude and the angle $(500\angle + 45°)$, and the rectangular form used trignometry to solve for either the north or east unknown quantity. When using the trignometric or circular method, both of the unknown quantities (north and east) are found after one mathematical calculation. In application, the magnitude of the vector is multiplied by the cosine function of the angle and sine function of the angle and is written as: magnitude $(\text{Cos } \Theta \pm j \text{ Sin } \Theta)$. All of this is pulled together in Figure 2-17 since 500 (Cos 45° + j Sin 45°) equals 500 (0.707 + j 0.707). The resulting answer tells us both the eastern and northern unit of measurement. All of this is told in the trigonometric notation 353.5 + j353.5.

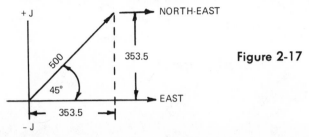

Figure 2-17

The same problem could have also been solved graphically if drawn to scale; then the desired answer could be measured. Graphic solutions, however, do not give as much accuracy because of human drawing errors or measuring accuracy. For this reason vector problems are normally solved using trigonometry.

2.5-6   Vector Addition

Vectors may be directly added or subtracted. To illustrate the addition process; suppose that we had two different magnitudes at different angles. Vector "A" is 3 at 40° and vector "B" is 2 at 20°, as illustrated in Figure 2-18.

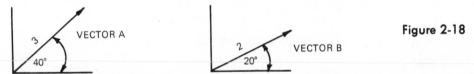

Figure 2-18

One good way of providing a rough check for the answer is first to draw vector "A" and then draw vector "B" starting at the end of vector "A." This is called a line vector drawing and is shown in Figure 2-19. The answer (hypotenuse) is the dotted line

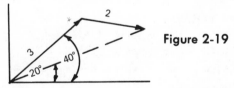

Figure 2-19

extending from point x to point y. If each vector were drawn to scale, the hypotenuse probably would measure roughly about 4.8 to 4.9. A protractor used to measure the new angle would show that an angle of about 30 degrees existed. The resultant vector, according to our line vector drawing, should be about 4.8 or 4.9 at 30 degrees. We shall, for accuracy's sake, solve this vector addition problem using the following methods.

For greater accuracy the same problem can be solved using trigonometry after first changing both vectors 3∠40° and 2∠20° to the rectangular form. To change from the polar form to the rectangular form, simply draw the vector as shown in Figures 2-20A and 2-20B. After doing this, the solutions are as follows:

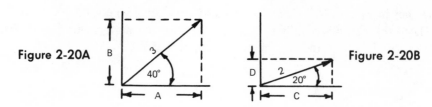

Figure 2-20A

Figure 2-20B

First find "B"

$$\sin 40° = \frac{\text{opp}}{\text{hyp}} = \frac{B}{3}$$

$$0.6428 = \frac{B}{3}$$

$$(3)(0.6428) = B$$
$$B = 1.9284$$

First find "D"

$$\sin 20° = \frac{\text{opp}}{\text{hyp}} = \frac{D}{2}$$

$$0.3420 = \frac{D}{2}$$

$$(2)(0.3420) = D$$
$$D = 0.684$$

Then find "A"

$$\cos 40° = \frac{adj}{\text{hyp}} = \frac{A}{3}$$

$$0.7660 = \frac{A}{3}$$

$$(3)(0.7660) = A$$
$$A = 2.298$$

Then find "C"

$$\cos 20° = \frac{adj}{\text{hyp}} = \frac{C}{2}$$

$$0.9397 = \frac{C}{2}$$

$$(2)(0.9397) = C$$
$$C = 1.8794$$

If we now take this data and add the rectangular forms representing both vectors as shown below, we can continue by applying these values to an equivalent triangle, thereby completing the problem.

Rectangular form addition:

$$2.2980 + j\,1.9284 \qquad \text{for vector } 3\angle40°$$
$$\underline{1.8794 + j\,0.6840} \qquad \text{for vector } 2\angle20°$$
$$4.1774 + j\,2.6124 \qquad \text{vectors sum in rectangular form}$$

Since we have determined the values for the individual vector problems we can redraw the triangles as illustrated in Figures 2-21A and 2-21B.

Figure 2-21A

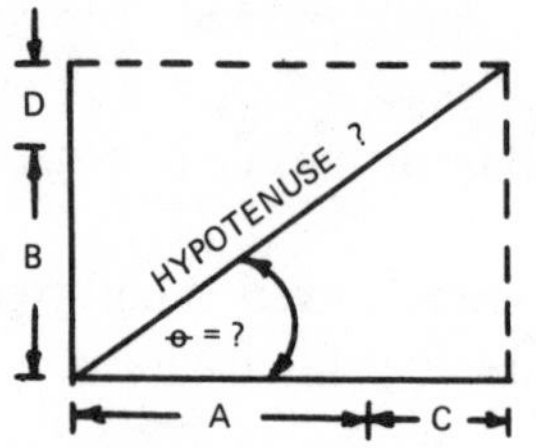

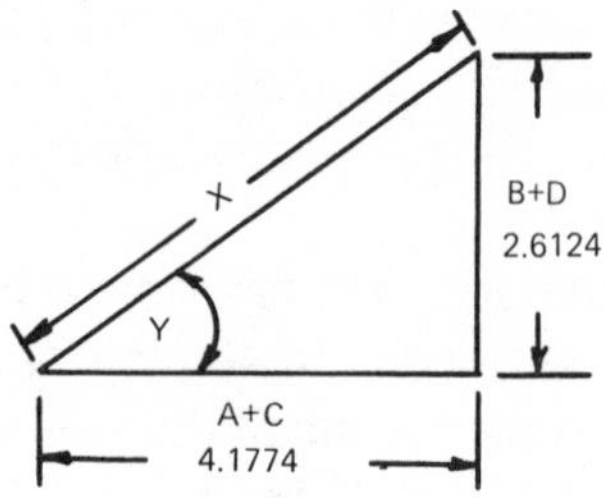

Figure 2-21B

Solving for angle Y

$$\text{Tan } y = \frac{\text{opp}}{\text{adj}} = \frac{2.6124}{4.1774}$$

$$\text{Tan } y = 0.6253$$

$$y = 32.0°$$

Solving for X

$$\text{Sin } y = \frac{\text{opp}}{\text{hyp}} = \frac{2.6124}{x}$$

$$0.5299 = \frac{2.6124}{x}$$

$$x = \frac{2.6124}{0.5299}$$

$$x = 4.93$$

The resulting vector is $4.93 \angle + 32°$ which is somewhat more accurate than the previous 4.8 or 4.9 at 30° obtained in the line vector method previous used.

2.5-7   Capacitive Reactance Vectors

The formula for capacitive reactance is:

$$X_C = \frac{1}{2 \pi fC}$$

X_c = capacitive reactance (ohms)
π = 3.14
f = frequency in hertz
C = capacitance in farads

Since current flow is strong and then reduces while the capacitor is building up a charge, the effect of capacitive reactance (X_c) is delayed. Because of this delay, the vectorial illustrations shown in Figure 2-22 are drawn 90° clockwise with respect to pure resistance.

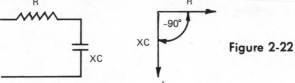

Figure 2-22

2.5-8   Inductive Reactance Vectors

The formula for inductive reactance is:

$$X_L = 2 \pi fL$$

X_L = inductive reactance (ohms)
π = 3.14
f = frequency in hertz
L = inductance in henries

Inductive reactance (X_L) is high initially, so current is slow building up; hence X_L leads R, and its vectorial representation is drawn 90° before pure resistance, as indicated in Figure 2.23.

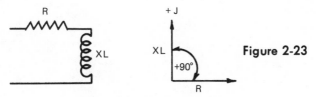

Figure 2-23

2.5-9   Impedance Vectors

Impedance is a vectorial combination of reactance and resistance; the total opposition to current flow in an AC circuit. In addition to the impedance (Z) magnitude, its phase angle (Θ) must be considered. Figures 2-24A and 2-24B illustrate the X_L and X_C impedance vectors.

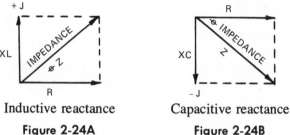

Inductive reactance   Capacitive reactance

Figure 2-24A   **Figure 2-24B**

Impedance calculations can be accomplished graphically, trigonometrically, or algebraically. Trigonometrically is best if both the size of impedance and the value of Θ are desired. Algebraically, using the Pythagorean Theorem, will give the ohmic value for impedance but not the phase angle Θ.

2.5-10   Resonance Vectors

When both forms of reactance are present, the circuit can become resonant. For any RLC circuit this condition occurs at only one specific frequency.

The resonance formula is:

$$F = \frac{1}{2\pi\sqrt{LC}}$$

where: F_r is the resonant frequency
L is in henries
C is in farads
$\pi = 3.14$

At this specific frequency (resonance) X_C and X_L cancel completely. The circuit, however, does contain both inductance and capacitance, acts as if it were only resistive, as indicated in Figure 2-25.

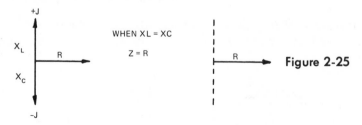

WHEN XL = XC

Z = R

Figure 2-25

XL and XC as vectors are 180° apart and cancel each other. This happens at resonance. However, when the circuit is not at the resonance frequency, the circuit may act either capacitive or inductive depending upon which value of reactance is greater. When one reactive component is larger than the other, impedance (Z) is computed by subtracting the two reactive values. This is illustrated by Figure 2-26.

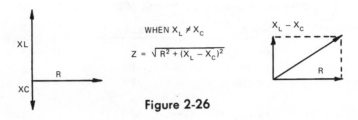

Figure 2-26

2.6   LOGARITHMS

Logarithms, once understood, provide shortcuts for lengthy electronic mathematical operations. Calculations involving exponents, powers, and roots can be accomplished more easily using logarithms than the arithmetical processes normally used.

A logarithm of any given number by definition is an exponent which indicates the power the given base must be raised to equal the given number.

There are two logarithm systems commonly used in electronic calculations. The first is called the *common or Briggsian* system of logarithms. The number 10 is used as the base for the common logarithm system. The second logarithm system is called the *natural or Napierian* system. The Greek letter epsilon (ϵ), which approximately equals 2.718, is used for the base for the natural logarithm system.

It might be noted that any positive number, with the exception of unity, is or can be used for a base in a logarithm system. The particular system used for calculations is indicated in the equation by the subscript which follows the abbreviation. For example, the equation X = Log 10N has the number ten for its base; the equation X = Log 2N has the number two for the base; the equation X = Log ϵN uses the Greek letter Epsilon as its base; and the equation X = Log bN uses the letter "b" for its base. If for some reason the base is not indicated, as in the equation X = Log N, the base of ten is understood, but if the equation is written in this manner, X = 1n N then the base is understood to be the Greek letter epsilon.

2.6-1   Common or Briggsian Logarithms

Logarithms consist of two basic parts, a whole number termed the *characteristic* and a decimal number called the *mantissa*. The mantissa is normally the only part recorded in logarithm tables and oftentimes even the decimal points are omitted. The characteristics are determined by inspection of the number as you will soon discover when applying the following rules and facts.

2.6-1A   POSITIVE CHARACTERISTICS

The characteristic of a logarithm for any number greater than the whole number one is always positive. The value of the characteristic is equal to one digit less than the number of digits required to write the whole number. For example, the characteristics for the following logarithmic equations are as follows:

Number		Characteristic
Log 1	=	0
Log 10	=	1
Log 100	=	2
Log 1,000	=	3
Log 10,000	=	4
Log 100,000	=	5
Log 1,000,000	=	6

Since the logarithm for 1 is 0 and the logarithm for 10 is 1, it should be evident that the logarithm for numbers between 1 and 10 will be somewhere between 0 and 1. In the same manner, a number between 10 and 100 would have a logarithm between 1 and 2.

2.6-1B   NEGATIVE CHARACTERISTICS

The characteristic for any number less than the whole number one is always negative. The value of the characteristic is equal to one digit more than the number of zeros counted between the decimal point and the first significant digit which is not a zero. This is illustrated in the following logarithmic equations.

Number		Characteristic
Log 0.1	=	−1
Log 0.01	=	−2
Log 0.001	=	−3
Log 0.0001	=	−4
Log 0.00001	=	−5
Log 0.000001	=	−6

For convenience, negative logarithms are usually transformed into an equivalent positive characteristic form. Characteristics of logarithms for any number less than one are simply written as differences between a new number representing the characteristic and the numbers of zeros between the decimal point and first significant digit. A negative number written after the mantissa will, after subtracting it from the new characteristic number, equal the negative logarithm. In order to understand this process, follow this listed data carefully.

Number	Characteristic		Transformed logarithm
Log 0.1	−1	=	9.0000−10
Log 0.01	−2	=	8.0000−10
Log 0.001	−3	=	7.0000−10

Number	Characteristic		Transformed logarithm
Log 0.000 0001	-7	$=$	$3.0000-10$
Log 0.000 000 000 000 001	-15	$=$	$5.0000-20$

Whenever nine or less zeros exist between the decimal point and first significant digit, a -10 is written after the mantissa. In like manner, whenever there are 10 to 19 zeros between the two references, a -20 is written.

2.6-1C   LOGARITHM MANTISSAS

Logarithmic tables give in the body of the tables the mantissa of logarithms for numbers ranging from 1 to somewhere in the hundreds or perhaps even thousands depending upon which tables you have. The number of places listed in these tables also may vary from four-place, five-place or higher, depending upon the size of the table.

When locating a mantissa for a number, the decimal point in the number can be disregarded because the mantissa is the same regardless of the decimal point position in the original number. This is demonstrated as follows:

$$Log\ 500.00 = 2.6990$$
$$Log\ \ 50.00 = 1.6990$$
$$Log\ \ \ \ 5.00 = 0.6990$$
$$Log\ \ \ \ 0.50 = 9.6990 - 10$$
$$Log\ \ \ \ 0.05 = 8.6990 - 10$$

If you looked at a common logarithm table for the log of 500, you would see the mantissa of 0.6990. Then, after you determined the characteristic for the number 500, you could say that $\log_{10} 500$ equals 2.6990. The following two equations show the exponential form and the logarithmic form for the same number.

$$10^{2.6990} = 500 \qquad\qquad 2.6990 = Log_{10}\ 500$$

(Exponential form)   (Logarithmic form)

2.6-2   How to Use Logarithmic Tables

In order to find the logarithmic for a particular number, you will need a common logarithm table. Listed across the top will be a column listing numbers normally marked with the capital letter "N" and digits ranging from 0 to 9 which are used to describe the particular number. The numbers begin with the number 10 and continue into the hundreds, or thousands. We will use a small part of a four-place logarithmic table to develop the remaining data on table usage.

COMMON LOGARITHM OF NUMBERS EXCERPT

N	0	1	2	3	4	5	6	7	8	9
69	8388	8395	8401	8407	8414	8420	8426	8432	8439	8445
70	8451	8457	8463	8470	8476	8482	8488	8494	8500	8506

We will demonstrate how to find the mantissa for the number 69.5 from the table on page 139 by following two simple steps. These steps are:

1. Find the first two digits of the given number under the "N" column.
2. Follow this line to the column headed by the third digit of the given number and record the mantissa for the three-digit number.

Since the characteristic is positive and equals 1, the common logarithm value for 69.5 is 1.8420. The following logarithms can also be obtained from the previous logarithm table.

$$
\begin{aligned}
\text{Log } 699 \quad &= 2.8445 \\
\text{Log } 7.05 \quad &= 0.8482 \\
\text{Log } 0.70 \quad &= 9.8451-10 \\
\text{Log } 0.007 &= 7.8451-10
\end{aligned}
$$

2.6-3   Antilogarithms

Finding an antilogarithm is just opposite to the process of finding a logarithm. In other words, finding the antilogarithm is finding the number which corresponds to a given logarithm. The antilogarithm process is indicated whenever the term antilog or $\log^{-1}$ is written.

The tables containing the mantissas are used when finding the decimal part of the logarithm and the characteristic of the logarithm tells us the position of the decimal point in the resulting number. The rules for characteristic-decimal point relationships are as follows:

1. If the characteristic of the logarithm is positive, count from the left of the resulting number one more digit then expressed by the characteristic.
2. If the characteristic of the logarithm is negative such as $9.8451-10$, the resulting number will have as many zeros minus one zero between the decimal point and the first significant digit as the number 10 following the mantissa in our case exceeds the characteristic of 9.

To apply these rules, examine what is happening to the following examples:

$$
\begin{aligned}
\text{Antilog} \quad 2.8445 \quad &= 699 \text{ (rule 1)} \\
\text{Antilog} \quad 0.8482 \quad &= 7.05 \text{ (rule 1)} \\
\text{Antilog} \quad 9.8451-10 &= 0.70 \text{ (rule 2)} \\
\text{Log}^{-1} \quad 7.8451-10 &= 0.007 \text{ (rule 2)}
\end{aligned}
$$

2.6-4   Logarithmic Multiplication

Since logarithms are exponents of 10, multiplication is accomplished in the same general way as exponents of algebra. Since we add the exponents in algebra when the bases are identical, we can say the logarithm of the product of two or more numbers is equal to the sum of their logarithms. To repeat this in formula form, the results are:

$$
\begin{aligned}
10^a \, 10^b &= 10^{a+b} \text{ (algebraic form)} \\
\text{Log } MN &= \text{Log } M + \text{Log } N \text{ (logarithmic form)}
\end{aligned}
$$

The logarithmic form above may seem misleading unless you remember that the base of 10 is understood when written as simply a log of a number. For application of multiplication, we will multiply the logarithms of 50 and 500 in the following manner:

$$(50)(500) = \text{antilog of log } 50 + \log 500$$

Since Log 50   = 1.6990
and Log 500   = 2.6990

Then the product is their sum.   1.6990
            + 2.6990
               4.3980

Consulting the log table for the number whose mantissa is 0.3980 we find the number 25. In order to determine the decimal point position we must consider the characteristic. Recall rule (1) outlined in 2.6-1A for determining the characteristic, in particular that this number was one less than the number of digits to the left of the decimal point. Because of this we must add one more digit to the present characteristic of 4. This new total of 5 tells us that the number 25 must have five digits to the left of the decimal point. This new number is 25000 as shown by:

$$(50)(500) \text{ or } 25000 = \text{antilog of log } 50 + \log 500$$

2.6-5   Logarithmic Division

Division by logarithms is accomplished by subtracting the log of the divisor from the log of the dividend. The simplified formula indicating this process is:

$$\text{Log} \frac{N}{M} = \text{Log } N - \text{Log } M$$

For an example, we will divide 25000 into 500.

$$\frac{500}{25000} = \text{antilog of log } 500 - \log 25000$$

Since log 500 = 2.6990
and log 25000 = 4.3979

Then the quotient is their difference: 2.6990
                          −4.3979
                        8.3011 − 10

After looking at the log table for a number whose mantissa is 0.3011 we find the number 20. This also does not determine the position of the decimal point. Because of rule #2, outlined in 2-6-1B, a number whose characteristic is negative has one less zero to the right of the decimal point than the value of the negative characteristic. Therefore the characteristic is 8− 10 = −2 minus the one, so there is one zero after the decimal and before the number 20 obtained for the quotient. The final answer is 0.02 and is expressed as:

$$\frac{500}{25000} = 0.02 = \text{antilog of log } 500 - \log 25000$$

2.6-6   Raising Powers of Logarithms

Raising any number to a power using logarithms is done by multiplying the logarithm of the number by the exponent of the power. This concept is represented by the following formula:

$$N^X = \text{antilog } (X \log N)$$

To show this formula in action we will raise the number 25 to the third power. The process using logarithms is shown below:

$$(25)^3 = \text{antilog of } 3 \ (\log 25)$$

$$\text{Since Log } 25 = 1.3979$$

$$\text{Then: the result is } 3(1.3979) \text{ or } 4.1937$$

When we look for the number whose mantissa is 0.1937 we find the value of 15625 (the last two digits being found by interpolation). Since the mantissa numbers are 1931 and 1959 and because the 1937 value is about ¼ of the difference a value of 0.25 is used. Following the format given in rule #2, outlined in 2.6-1B, the characteristic of 4 provides five digits to the left of the decimal point or 15,625.

2.6-7   Extracting Roots of Logarithms

The logarithm of the root for any number is the logarithm of the number divided by the power of the root. The equation illustrating this process is as follows:

$$\text{Log } ^X\!\sqrt{N} = \frac{\text{Log } N}{X}$$

We will find the antilog of the cube root of the number 125. The operations necessary are as follows:

$$^3\sqrt{125} = \text{antilog of } \frac{\text{Log } 125}{3}$$

$$\text{Since Log } 125 = 2.0969$$

$$\text{Then: } \frac{\text{Log } 125}{3} = 0.6989$$

Since the characteristic of zero provides one digit to the left of the decimal point, the cube root of 125 is 5. This is logical because $5 \times 5 \times 5$ equals 125.

2.7   ADVANCED MATHEMATICS ANALYSIS

It seems that limits exist for the usefulness of specialized areas of mathematics. For example, in elementary mathematics, the basic limits seem to be adding, subtracting, and dividing; algebra confines itself to solving unknown equations; while geometry and trigonometry are contained within circles. Now, in all of these areas of specialized math, a dependency exists, between one and the other. In this portion of

Chapter 2, combinations of these specialized areas of mathematics result in what might be called analytic geometry; and calculus will be outlined. In keeping with the "thumbnail sketch' approach found within the covers of this book, facts, rules and other pertinent data dealing with advanced mathematics will be given in eye-opening—brain-reaching outlines.

2.7-1   Functions

Mathematics in general uses two kinds of symbols to represent numbers or numerical quantities. These are called variables and constants. In higher or advanced mathematics, two variables can have a relationship between each other such that for each set of values given to one, there results one or more definite values for the other. When this exists, it is said one is a function of the other. The first variable is called the dependent function or variable, while the second variable is termed the independent variable or argument.

If the letter "Y" is a function of the letter "X" and for each value of "X" there is a value or values for "Y," then this can be written in equation or functional notation form illustrating the same. To indicate that "Y" is a function of "X" we write $Y = f(X)$. This equation is read "Y" equals f of "X" or "Y" equals a function of "X." The symbol $f(X)$ represents any mathematical expression involving only the variable "X" or any quantity that is a function of "X."

If the variable in a functional notation equation is replaced with a number as in the following examples, it indicates that the variable is to be made equal to that number. For example, in the equation $y = x^2 - 2$, the equation has different answers when solved for different functions.

$$\text{(Given Equation) } y = x^2 - 2$$

$$f(x) \text{ function } = x^2 - 2$$
$$f(0) \text{ function } = 0 - 2 = -2$$
$$f(1) \text{ function } = 1 - 2 = -1$$
$$f(3) \text{ function } = 9 - 2 = 7$$
$$f(e) \text{ function } = (e)^2 - 2$$

Inverse functions are also possible when you regard "X" as the dependent variable and "Y" as the independent variable. Thus, the function of $y = x^2 - 2$ and $x = \sqrt{y + 2}$ are inverse functions.

2.7-2   Increments

An increment for mathematics is defined as a variable which changes from one numerical value to another. The increment of the variable is determined by subtracting the first value of the variable from the second value. With this in mind, if current (I) is 5.6ma for one reading and then 5.8ma, the increment for "I" is 0.2. Increments are also taken when variables decrease, such as 5.8 for the first reading to 5.6ma for the second. The increment for "I" in this case would be -0.2.

In practice, the word increment is symbolized by the Greek letter delta symbol (Δ). If this symbol is prefixed to the letter "X" representing the variable, then ΔX

denotes an increment of X. When a variable is a function of another variable the following truths exist. We will use "x" as a function of "y" or x = f(y) for illustration purposes. If you let the first value of the function, y in our case, equal y_1 and the second value equal y_2, then $\Delta y = y_2 - y_1$ or $y_2 = y_1 + \Delta y$. Because "x" is a function of "y" and since "y" is taken on the increment Δy, then "x" will also have a corresponding increment. The values for "x" will be represented by x_1 and x_2. All of this means that $\Delta x = x_2 - x_1$ or $\Delta x = f(y_2) - f(y_1)$ or $\Delta x = f(y_1 + \Delta y) - f(y_1)$.

The solution for Δx in the equation $x = y^2 - 6y$ when $y = 2$ and $\Delta y = 3$ is obtained in the following manner:

1. Substitute 2 for "y" in the equation $x = y^2 - 6y$ solving for $f(x_1)$ when $y_1 = 2$.

$$x_1 = f(y_1) = 2^2 - 6(2)$$
$$x_1 = -8$$

2. Find the "value of y_2 by adding the values given for y and Δy."
$$y_2 = 2 + 3 = 5$$

3. Solve for $f(x_2)$ when $y_2 = 5$ in the equation $x = y^2 - 6y$.

$$x_2 = f(y_1 + \Delta y)$$
$$x_2 = (5)^2 - 6(5)$$
$$x_2 = -5$$

4. Substitute x_2 representing $f(y_1 + \Delta y)$ and x_1 representing $f(y\ 1)$ into the following equation solving for Δx.

$$\Delta x = f(y + \Delta y) - f(y) \qquad \text{or:} \quad \Delta x = x_2 - x_1$$
$$\Delta x = -5 - (-8) \qquad\qquad\qquad \Delta x = -5 - (-8)$$
$$\Delta x = +3 \qquad\qquad\qquad\qquad\quad \Delta x = +3$$

2.7-3   Slope

There are three words used in higher mathematics which might seem confusing even to the experienced. These three are: inclination, angle, and slope. The one basic thing they have in common is that they all define lines. An *inclination* is the angle which exists between the given line and the horizontal or "X"-axis. The *angle* existing between two directed lines is the angle whose sides extend from the angle's vertex. The *slope* of a line is a tangent function of its inclination.

The slope provides a convenient way of telling line direction. The slope normally denoted by a small letter "m" is written as: m = tan θ, where theta (θ) or some other symbol represents the angle for the line. The slope may be any real positive or negative number. If the angle is less than 90 degrees, the slope is positive, while if the angle is more than 90 degrees, the slope is negative.

2.7-3A   SLOPE FORMULA

The slope is considered to be the rate of change relationships between the "Y" axis values and the "X" axis values as shown in Figure 2-27.

The slope, considered to be the rate of change of the ordinate (Y-coordinate of

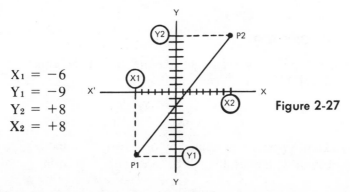

$$X_1 = -6$$
$$Y_1 = -9$$
$$Y_2 = +8$$
$$X_2 = +8$$

Figure 2-27

the point) with respect to the abscissa (X-coordinate of the point), can be written using increment notation. Thus $m = \dfrac{\Delta Y}{\Delta X}$ where ΔY is the increment of $Y_2 - Y_1$ and ΔX is the increment of $X_2 - X_1$. The slope of the line passing through points P_1 or X_1, and P_2 or X_2, Y_2 may also be expressed as:

$$m = \frac{Y_2 - Y_1}{X_2 - X_1} = \frac{8 - (-9)}{8 - (-6)} = \frac{17}{14}$$

$$m = \frac{17 \text{ representing ``Y''}}{14 \text{ representing ``X''}}$$

2.7-3B   POINT SLOPE FORM

The point slope form provides an easy way for constructing a line which passes through a given point at a given slope. Therefore, an equation for a straight line can be recorded if you want the slope and one point on that line. The equation illustrating the point slope form is:

$$X - X_1 = m (Y - Y_1)$$

X and Y represent the unknown coordinates.

2.7-3C   SLOPE-INTERCEPT FORM

You can easily determine the slope and the "Y" intercept value for a straight line using the slope-intercept equation form. The "X" intercept is the "X" coordinate of the point of intersection of the line with the "X" axis, while the "Y" intercept is the "Y" coordinate of the point of intersection of the line with the "Y" axis. The general slope-intercept formula form is:

$$Y = mX + b$$

The slope and "Y" intercept for the equation $5y - 25X - 30$ is automatically in the $Y = mX + b$ form upon completion of the following solution for "Y":

$$5Y - 25X = 30$$
$$5Y = 25X + 30$$
$$Y = \frac{25X + 30}{5}$$
$$Y = 5X + 6 \text{ (the } mX + b \text{ form)}$$

2.7-3D   TWO-POINT FORM

When two points of a straight line are known, an equation representing the two-point form can be written. The equation is:

$$\frac{Y-Y_1}{X-X_1} = \frac{Y_2-Y_1}{X_2-X_1}$$

after the two known points represented by their two "X" and "Y" coordinate values have been substituted, the results will be in the mX + b form.

2.7-3E   SERIES RC CIRCUIT APPLICATION

Instantaneous values for currents and voltages during specific periods of time can be solved quite easily when higher mathematic techniques are employed. Figure 2-28 illustrates a series resistive, capacitive circuit and its vectorial graph showing voltage-time change relationships. Our task will be to find the instantaneous current "i" value when the 2 microfarad capacitor has charged to 10 volts after 1 microsecond of time. Needless to say, i_c is the instantaneous capacitor current, e_c the instantaneous capacitor voltage, and t_o, t_1 the time limits.

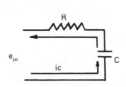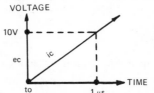

Figure 2-28

If the capacitor voltage is increasing as the time increases, an increment of $\dfrac{\Delta e}{\Delta t}$ showing a change in voltage with respect to time can be written. A formula representing the slope from showing these same effects could be written as:

$e_c = Kt$ where e_c is the capacitor voltage

K is the slope
t is the time in seconds.

We will use the equation $i_c = C\dfrac{\Delta e}{\Delta t}$ to determine the current flowing in the capacitive circuit after finding the slope also called the rate of change of voltage with respect to time. This is accomplished by substituting the values for e_c and t into the equation for e_c as follows:

Given: $e_c = 10$ volts

$\Delta t = 1$ microsecond $= 1 \times 10^{-6}$

Therefore: $e_c = Kt$

$$K = \frac{e_c}{t} = \frac{10}{1 \times 10^{-6}} = 10 \times 10^{+6}$$

$$K = 10 \times 10^{+6}$$

since the slope (K) is the same as $\dfrac{\Delta e_c}{\Delta t}$, then the values for the capacitor and $\dfrac{\Delta\ ec}{\Delta\ t}$ can be substituted

into the equation describing the current flow in a capacitive circuit. The results are:

$$ic = C\,\dfrac{\Delta\ ec}{\Delta\ t}\ \text{or CK}$$

$$ic = (2\times 10^{-6})\,(10\times 10^{+6}) = 20\ \text{amperes}$$

ic = 20 amperes. This is the instantaneous current flowing for 1 microsecond thereby causing an instantaneous voltage of 10 volts across the capacitor to exist.

2.7-4   Derivatives

In the examples defining their slopes as $\dfrac{\Delta\ Y}{\Delta\ X}$ in the previous 2.7-3 sections of this chapter, derivatives of the first variable with respect to the second variable were implied. The derivative of the first variable with respect to the second is implied whenever one variable is a function of a second function, and the increment of the first function is divided by the increment of the second, and when the limit of the quotient is as the second increment approaches zero. We will use "Y" as a function of "X" to show how this definition looks when in formula form. The formula is:

$$y = f'\ (x) = \underset{\Delta\ x\to 0}{\text{Limit}}\quad \dfrac{\Delta\ y}{\Delta\ x}\ \text{at x = x}_1$$

The f' simply means that the derivative of "y" with respect to "x" having the limit of $\dfrac{\Delta\ Y}{\Delta\ X}$ as Δ x approaches zero. The derivative process for "y" with respect to "x" is also denoted by:

$$\dfrac{dy}{dx}\ \text{or y' or f'(x)\ or}\ \dfrac{d}{d(x)}\ f(x)$$

The process used to find the derivatives of functions is called differentiation. The following generalized rules will help you differentiate one variable with respect to the other. We will use the y = f(x) form for the following:

1. Give the "x" variable an increment Δ x and substitute x + Δ x in the equation describing the variable "y".

2. Calculate the value for y + Δ y, expending the right-hand equation member.

3. Subtract "y" from y + Δ y, obtaining Δ y in terms of x and Δ x.

4. Divide Δ y by Δ x, obtaining the value for $\dfrac{\Delta\ y}{\Delta\ x}$.

5. Find the limit for the expression found in step 4 as Δ x approaches zero (Δ x → 0).

The result will be the value of $\dfrac{dy}{dx}$, y', f'(x) or $\dfrac{d\ f(x)}{d(x)}$, depending upon which notation for finding the derivative you employed.

For example, we will find the derivative of "y" for the equation $y = x^3 - 2x + 4$.

Rule 1: substitute x + Δ x for x.

$$y = x^3 - 2x + 4$$
$$y = (x + \Delta x)^3 - 2(x + \Delta x) + 4$$

Rule 2: Calculate the value for y + Δ y, expanding the right-hand equation member.

$$y + \Delta y = (x + \Delta x)^3 - 2(x + \Delta x) + 4$$
$$y + \Delta y = x^3 + 3x^2 \Delta x + 3x(\Delta x)^2 + (\Delta x)^3 - 2x - 2 \Delta x + 4$$

Rule 3: Subtract "y" from the left-hand equation member and the value for "y" (which equals $x^3 - 2x + 4$) from the right-hand equation member, resulting in:

$$\Delta y = 3x^2 \Delta x + 3x(\Delta x)^2 + (\Delta x)^3 - 2 \Delta x$$

Rule 4: divide Δ y by Δ x, obtaining:

$$\frac{\Delta y}{\Delta x} = 3x^2 + 3x(\Delta x) + (\Delta x)^2 - 2$$

Rule 5: Find the limit for the expression of step 4. Note that the second and third terms of the right-hand equation member approach zero if Δ x → 0 therefore:

$$\frac{dy}{dx} = 3X^2 - 2$$

2.7-4A   DIFFERENTIATION FORMULAS

Rather than deriving formulas by differentiation to show you their origin, etc, we shall give you the conditions and formulas generally used in differential calculus.

1. *Derivative of a constant*
   If the letter "c" is a constant and if y = c then:
   $$\frac{dy}{dx} = 0 \quad or \quad \frac{dc}{dx} = 0$$

2. *Derivative of a first-order variable*
   This simply means that a derivative of a variable with respect to itself is one. If y = x then:
   $$\frac{dy}{dx} = 1$$

3. *Derivative of a variable having an exponent*
   if $y = x^N$, the formula is:
   $$\frac{dy}{dx} = Nx^{N-1}$$

4. *Derivative of a sum*
   If "u" and "v" are functions of "x" and if y = u + v then the following formula results:
   $$\frac{dy}{dx} = \frac{du}{dx} + \frac{dv}{dx}$$

5. *Derivative of a product*
   When "u" and "v" are functions of "x" and if $y = uv$ then the formula used is:

$$\frac{dy}{dx} = u\frac{dv}{dx} + v\frac{du}{dx}$$

6. *Derivative of a quotient*
   If $y = \dfrac{u}{v}$ and "u" and "v" are functions of "x", the following formula is used to find the derivative:

$$\frac{dy}{dx} = \frac{v\dfrac{du}{dx} - u\dfrac{dv}{dx}}{(v)^2}$$

7. *Derivative of a function having an exponent*
   The following formula is used when $y = u^m$ if the letter "u" is a function of "x".

$$\frac{dy}{dx} = (mu)^{m-1}\frac{du}{dx}$$

8. *Derivative of higher orders*
   This is termed second derivative, third derivative, fourth derivative, etc. Since the derivative $\dfrac{dy}{dx}$ of the function "y" is in itself a function, its derivative is termed the second derivative. The second derivative for dy is illustrated below in four general ways.

$$\frac{d\,(dy)}{dx\,(dx)} = \frac{dy}{dx} = y'' = f'' = dx^2y$$

The third derivative for the same is:

$$y''' \text{ or } f''' = \frac{d^3y}{dx^3}$$

The nth derivative would be:

$$y^n = \frac{d^ny}{dx^n}$$

2.7-5   Integrals

Many mathematical processes essential for calculation are really inverse operations. Some of the mathematical processes illustrating inverse operations include addition-subtraction, multiplication-division, and logarithms and exponents. The inverse operations for differentiation in calculus is called integration.

When you integrate, you are actually finding the original function when the derivative of the function is known. An example written in symbolic form is: $\int f'(x)dx = f(x)$. This expression is read as the integral of the first derivative of "y" with respect to "x" which equals the function of "x".

Circuitry having a resistor and a capacitor form a differentiator circuit or an integrator circuit depending upon which component the output is taken across. Figure 2-29 illustrates these two particular circuits.

Electronically speaking, a square wave when applied to the input or an integrator circuit will produce a triangular waveform and is expressed mathematically by using the integration sign. When a triangular wave is applied to the input of a differentiator circuit, the output wave form will resemble a square wave and is expressed mathematically as a differentiation. Hence, these two circuits perform inverse operations in wave forming circuits.

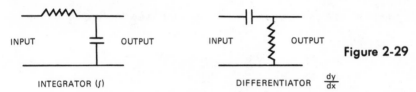

INPUT   OUTPUT   INPUT   OUTPUT

INTEGRATOR ($\int$)   DIFFERENTIATOR $\frac{dy}{dx}$

Figure 2-29

The voltage developed across either capacitor in Figure 2-29 can be expressed mathematically, but before we show you the formula, the following information will prove helpful. Capacitance symbolized by the letter "C" is the factor by which the voltage (V) measured across the capacitor is multiplied, thereby determining the charge (Q) on the capacitor. This is expressed as:

$$Q = CV \text{ where "Q" is in coulombs}$$
$$\text{"C" is in farads}$$
$$\text{"V" is in volts}$$

The derivative of the charge, (Q), with respect to time (t) describes the charging current flow for the capacitor. This is written as:

$$\frac{dq}{dt} = i = C\,\frac{dv}{dt} + \frac{dc}{dt}$$

Since capacitance is constant and not a function of time for most problems, dc/dt is zero and results in solving for current using:

$$i = C\,\frac{dv}{dt} \text{ where "i" is in amperes}$$

$$\frac{dv}{dt} \text{ is in volts/second}$$

The voltage (V) developed across the capacitor is determined when solving for "V" in the following process:

$$\text{since } i = C\,\frac{dv}{dt} \text{ then: } dv = \frac{1}{C}\,i\,d\,t$$

$$\text{solving for "V" results in: } \quad V = \frac{1}{C}\int_0^t i\,d\,t \pm E_0$$

where E_0 equals the voltage across the capacitor at zero time.

This equation represents the voltage which is measurable across the capacitor during a time interval from 0 seconds to the instant of time "t" seconds later. The number zero written at the bottom of the integration symbol is called the *lower limit* and the letter "t" written at the top is called the *upper limit*.

CHAPTER 3

Key Facts and Formulas for Electronic Problem Solving

The hand tools used in our profession have their intended primary purpose. You know the limitations, advantages, and disadvantages of tool usage, and therefore select which to use under certain circumstances. Knowledge of equal importance is selecting which fact, formulas, or concept can be relied upon when various conditions are present.

This chapter will enable you to find out which formula, fact, or concept can be used and relied upon, not only when used in DC or AC circuitry application, but also when other variables are important. The following main subject areas are arranged alphabetically while data pertinent to each subject has been logically, but not necessarily alphabetically, listed.

3.1 ADMITTANCE (AC APPLICATIONS ONLY)

Admittance is a measurement of ease with which AC current flows in any circuit. It is the reciprocal of impedance and is expressed in the following ways:

3.1-1 General

$$Y = \frac{1}{Z}$$

Y = admittance in mhos
Z = impedance in ohms

3.1-2 Series Circuit

$$Y = \frac{1}{\sqrt{R^2 + X^2}}$$

Y = admittance in mhos
R = Resistance in ohms
X = reactance in ohms

3.1-3 Parallel Circuit

$$Y = \sqrt{G^2 + B^2}$$

Y = admittance in mhos
G = conductance in mhos
B = susceptance in mhos

Note: The unit of mho describing conductance or admittance is replaced by the metric unit of siemens (S) whose formula is: $S = A/V$ or I/E

3.2   ALTERNATING CIRCUIT (AC APPLICATIONS ONLY)

Alternating circuit current is a flow of electrons whose movement direction periodically reverses. A complete cycle consists of two alternations: the positive alternation and the negative. Whenever more than one cycle exists, the term hertz is employed meaning cycles and involving frequency.

3.2-1   Terminology and Values

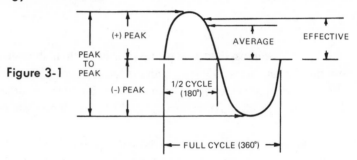

Figure 3-1

Peak to Peak—actual voltage swing, measured overall
Peak voltage—actual voltage swing, positive or negative
Effective voltage—value read on most meters (rms)
Average voltage—value sine wave is above or below the reference half the
time. One peak only.
Period—Time required for one complete cycle $t = \dfrac{1}{f}$ t = periods (Seconds
f = frequency (Hertz)

3.2-2   Conversion Formulas (AC Applications Only)

Peak value = $\dfrac{\text{peak to peak value}}{2}$
Peak value = 1.414 × effective (rms) value
Peak value = 1.57 × average value
Peak to peak value = 3.14 × average value
Peak to peak value = 2.828 × effective (rms) value
Peak to peak value = 2 × peak value
Effective or rms value = 0.707 × peak value
Effective or rms value = 1.11 × average value
Effective or rms value = 0.3535 × peak to peak value
Average value = 0.637 × peak value
Average value = 0.9 × effective value
Average value = 0.32 × peak to peak value

3.2-3   Instantaneous Voltage Specifics (AC Applications Only)

A sine wave is the waveform of an alternating current or voltage in which the amplitude varies as the sine of the phase angle between each instantaneous value and the last previous passage through zero from the negative to positive direction. When determining the actual value of the instantaneous voltage at any particular instant of time, simply multiply the sine function of the desired angle times the peak value of the voltage.

Less than 90 degrees

E inst. sine wave at 70° = ?

E inst. = sine 70° × E peak value

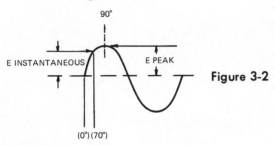

Figure 3-2

Greater than 90 degrees

*Use sign function of number of
 degree difference to closest
 zero voltage reference point.

E inst. sine wave at 112° = ?

E inst. = sine (180°− 120°) × peak value

E inst. = sine 68° × E peak value

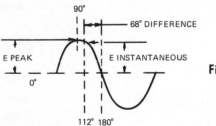

Figure 3-3

Greater than 180 degrees

E inst. sine wave at 262° = ?

E inst. = sine (262°− 180°) × E peak value

E inst. = sine 82° × E peak value

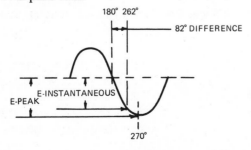

Figure 3-4

3.2-4　Ohms Law Variations (AC Application only)

$$E = IZ \qquad I = \frac{E}{Z}$$

$$E = IX \qquad I = \frac{E}{X}$$

$$E = \frac{P}{I \, Cos \, \Theta} \qquad Z = \frac{E}{I}$$

$$X = \frac{E}{I} \qquad Z = \frac{P}{I^2 \cos \Theta}$$

E = voltage in volts
I = current in amperes
X = inductive or capacitive reactance in ohms
P = power in watts
Θ = phase angle in degrees
Z = impedance in ohms

3.2-5　Phase Angles (AC Application Only)

The difference in degrees by which current leads voltage in a capacitive circuit or by which current lags voltage in an inductive circuit is termed phase angle. The formulas used to determine phase angles in series circuits are as follows:

$$\Theta = ARC \, \tan \, \frac{X_L}{R}$$

$$\Theta = ARC \, \tan \, \frac{X_C}{R}$$

Θ = angle of lead or lag in degrees
X_L = inductive reactance in ohms
X_C = capacitive reactance in ohms
R = non-reactive resistance in ohms

3.2-6　Power Factor (AC Application Only)

Power factor is a ratio between actual power of an alternating or pulsating current, measured by a wattmeter to an apparent power value as indicated by voltmeter and ammeter readings. Power factor is a cosine function of the phase angle between sinsusodial voltage and current.

Power factor also indicates loss in an inductor, capacitor, or insulator since it is a ratio of resistance and impedance. The formulas used to determine the power factor are:

$$PF = \frac{R}{Z}$$

$$PF = Cosine \, \Theta$$

$$PF = \frac{true \, power}{apparent \, power}$$

PF = power factor (percentage)
R = non-reactive resistance (ohms)
Z = impedance (ohms)
True power = EI cos Θ (watts)
Apparent power = EI (volt-amperes)

True power—Apparent power example:

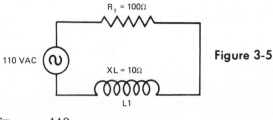

Figure 3-5

Since $I_T = \dfrac{E_T}{Z_T} = \dfrac{110}{110} = 1A$

Then apparent power is:

$PA = I^2Z$ OR $PA = IE$

$PA = (1)^2 110$ $PA = 1 \times 110$

$PA = 110$ watts $PA = 110$ watts

and true power is:

$P_T = I^2R$

$P_T = (1)^2\ 100$

$P_T = 100$ watts

Now that we know true power equals 100 watts and apparent power is 110 watts, the power factor can be determined.

$$PF = \frac{\text{true power}}{\text{apparent power}}$$

$$PF = \frac{100}{110} = 0.909$$

$$PF = 90.9\%$$

3.2-7 Angular Velocity (AC Application Only)

The Greek letter omega (ω) represents the angular velocity of alternating voltage or signal. Instead of writing 2π f in any formula requiring this particular trio the angular velocity symbol is employed because:

angular velocity (ω) = 2π f $\pi = 3.14$
                                                    f = frequency

3.2-8 Instantaneous Voltage (AC Application Only)

$e = E_{max} \sin \omega t$ e = instantaneous voltage at anytime (t)
                              E_{max} = maximum voltage
                                  ω = angular velocity (2π f)
                                  t = given time length

3.2-9   Potential Difference for a Charging Capacitor in "RC" Circuits

*Refer also to time constants (3.19)

$$e_C = E_A \left(1 - \epsilon \, \frac{-t}{RC} \right)$$

OR

$$e_C = E_A \left(1 - \xi^{-X} \right)$$

e_C	=	voltage across capacitor
E_A	=	applied voltage
ξ	=	base of Napierian equaling 2.718
t	=	desired time length
RC	=	circuitry change time constant
X	=	time in time constants $\left(\dfrac{t}{RC} \right)$

3.2-10   Potential Difference for a Discharging Capacitor in "RC" Circuits

*Refer also to time constants (3.19) (AC applications)

$$e_C = E_c \left(\xi \, \frac{-t}{RC} \right)$$

OR

$$e_c = E_c \left(\epsilon^{-X} \right)$$

e_c	=	voltage across capacitor
E_c	=	initial charge on capacitor
ξ	=	base of Napierian logarithm equaling 2.718
R_c	=	circuitry discharge time constant
X	=	time to time constants (t/R_c)
t	=	desired time length

3.2-11   Voltage Across a Resistor in "RC" Circuits (AC Applications)

*Refer also to time constants (3.19)

$$e_R = E_A \left(\xi^{-t/RC} \right)$$

e_R	=	voltage across resistor
E_A	=	applied voltage
ξ	=	Napierian log base (2.718)
t	=	any given time length (seconds)
R_c	=	time constant (seconds)

3.2-12   Voltage Across a Coil in "LR" Circuits (AC Applications)

*Refer also to time constants (3.19)

$$e_L = E_A \left(\xi^{-tR/L} \right)$$

e_L	=	voltage across coil.
E_A	=	applied voltage
t	=	any given time length (seconds)
R	=	circuitry resistance (ohms)
L	=	circuitry inductance (Henrys)

3.2-13   Voltage Across a Resistor in "LR" Circuits (AC Applications)

*Refer also to time constants (3.19)

$$e_R = E_A (1- \xi^{-tR/L})$$

e_R = voltage across resistor
E_A = applied voltage
ξ = Napierian log base (2.718)
R = circuitry resistance (ohms)
L = circuitry inductance (Henrys)
t = any given time length (seconds)

3.2-14   Instantaneous Current (AC Applications)

$$i = I_{max} \sin \omega t$$

i = instantaneous current at any time (t)
I_{max} = maximum current (peak)
$\omega = 2\pi f$

3.2-15   Instantaneous Current in LR Circuit (AC Applications)

*Refer also to time constants (3.19)

$$i = \frac{E_A}{R} (1-\xi^{-tR/L})$$

i = instantaneous current
E_A = applied voltage
R = circuitry resistance (ohms)
L = circuitry inductance (ohms)
t = any given time interval
ξ = Napierian log base (2.718)

3.2-16   Instantaneous Current in RC Circuits (AC Applications)

*Refer also to time constants (3.19)

$$i = \frac{E_A}{R} (\xi^{-t/RC})$$

i = instantaneous current
E_A = applied voltage
ξ = Napierian log base (2.718)
t = value of given time
R_C = circuitry time constant

3.2-17   Transit Time Phase Difference (AC Applications)

$$\propto = 2\pi f t$$

OR

$$\propto = \omega t$$

$\propto$ = transit time phase difference
π = 3.14
f = frequency
t = value of given time
ω = $2\pi f$

3.3   ANTENNAS (AC APPLICATIONS ONLY)

It seems that antenna design and reference data are based upon the fundamental single wire antenna type whose physical length is approximately equal to one-half its transmitting wavelength. The following formulas are listed for quick reference for antenna applications.

SINGLE WIRE ANTENNA FORMULAS:

3.3-1   Half Wavelength in Feet (in Space)

$$L = \frac{492}{f}$$

L = length in feet
f = frequency in megahertz

3.3-2   Half Wave-Actual Length in Feet

$$L = \frac{492 \times 0.95}{f}$$

OR

$$L = \frac{468}{f}$$

L = length in feet
f = frequency in megahertz
0.95 = resonant antenna length factor

3.3-3   Physical Length of Single-Wire (One Wavelength)

$$L = \frac{984 \times V}{f}$$

L = physical length in feet
V = velocity factor for particular transmission line used
f = frequency in megahertz

3.3-4   Electrical Length of Single Wire

$$\lambda = \frac{984,000}{f}$$

f = frequency in kilohertz
λ = wavelength in feet
(Electrical length is longer than actual length)

3.3-5   Long Wire Antenna Length

$$L = \frac{492(N-0.05)}{f}$$

L = length in feet
N = number of halfwaves on antenna
f = frequency in megahertz

3.3-6   Quarter-Wave Antenna Length

$$L = \frac{246(V)}{f}$$

L = length in feet
f = Frequency in megahertz
V = velocity factor of transmission line used

3.3-7   Antenna Radiation Resistance

$$R = \frac{P}{I^2}$$

R = radiation resistance
P = radiated power
I^2 = effective current (maximum current point on antenna)

3.3-8   Antenna Efficiency

$$EFF = \frac{Rr \times 100}{Rr + Rw + RL}$$

EFF = antenna efficiency
Rr = radiation resistance
Rw = effective wire resistance
RL = equivalent resistance (other losses)

3.3-9   Total Number of Antenna Lobes

$$LT = 2N$$

LT = total number of lobes
N = number of half-wave lengths on antenna

3.4   CAPACITORS (AC AND DC APPLICATIONS)

The one component used in electronics which has the ability "to pass alternating current or pulsating direct current voltages and blocks unchanging or direct current voltages" is the capacitor. The capacitor can and does store energy in the form of an electric field which exists between its plates when voltage is applied. The following rules, facts, and formulas are those we rely upon daily in the electronics profession.

3.4-1   Total Capacitance for Two series Capacitors (AC and DC Applications)

Figure 3-6

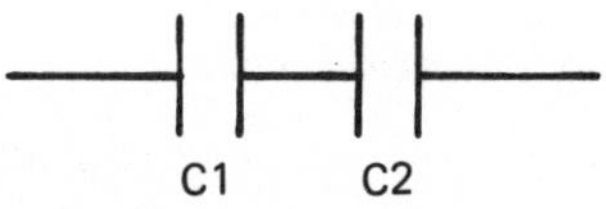

$$CT = \frac{C1 \times C2}{C1 + C2}$$

C1   C2

*The same unit of capacitance must be used; microfarads, micro-micro-farad (now pF), etc.

3.4-2   Total Capacitance for Any Number of Capacitors in Series (AC and DC Applications)

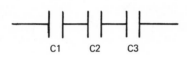

$$C_T = \cfrac{1}{\cfrac{1}{C1} + \cfrac{1}{C2} + \cfrac{1}{C3} + \ldots}$$

Figure 3-7

*Same unit of capacitance must be used.

3.4-3   Voltage Across Capacitor When Capacitors Are in Series (DC Application Only)

$$E_{CX} = \frac{E_A\ (C_T)}{C_X}$$

E_{CX} = desired capacitor's voltage

E_A = applied DC voltage

C_T = total circuitry capacitance

C_X = desired capacitor's value

*E_{CX} is inversely proportional to its capacitive value

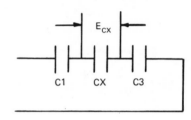

Figure 3-8

3.4-4   Total Capacitance for Capacitors in Parallel (AC and DC Applications)

$C_T = C_1 + C_2 + C_3 + \ldots$

C_T = total capacitance

C_1-C_3 = individual capacitances

*Largest voltage safely applied equals smallest voltage rating of the capacitors.

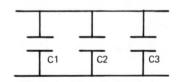

Figure 3-9

3.4-5   Capacitor's Plate Capacitance (AC and DC Applications)

When area of one plate (A) and dielectric thickness (D) are both in inches, use the first formula.

$$C = 0.2235\ \frac{(KA)(N-1)}{D}$$

C = plate capacitance in picofarads

K = dielectric constant

A = area of one plate (Sq. inches)

D = dielectric thickness (inches)

N = number of plates

When area of one plate (A) and dielectric thickness (D) are both in centimeters, use this formula:

$$C = 0.0884 \frac{(KA)(N-1)}{D}$$

C = plate capacitance in picofarads
K = dielectric constant
A = area of one plate (sq. centimeters)
D = dielectric thickness (centimeters)
N = number of plates

3.4-6   Energy Stored in a Capacitor (DC Applications)

$$W = \frac{CE^2}{2}$$

W = energy in joules (watt-seconds)
C = capacitance in farads
E = applied voltage

3.4-7   Charge Stored in a capacitor (DC Applications)

$$Q = CE_c$$

Q   = charge in coulombs
C   = capacitance in farads
E_c = voltage across capacitor

3.4-8   Total Charge on Capacitors in Series (DC Applications)

$$Q_T = Q_1 = Q_2 = Q_3$$ Q = charge in coulombs

3.4-9   Total Charge on Capacitors in Parallel (DC Application)

$$Q_T = Q_1 = Q_2 + Q_3 + \ldots$$ Q = charge in coulombs

3.4-10   Quality factor, Q, or Figure of Merit for Capacitors (AC Applications)

For single capacitor:

$$Q = \frac{X_c}{R}$$

Q   = reactance-resistance ratio
X_c = capacitive reactance in ohms
R   = Resistance which acts in series
       with capacitor (in ohms)

For a capacitor in series with a resistor:

$$Q = \frac{1}{2\pi f RC}$$

OR

$$Q = \frac{1}{\omega RC}$$

Q   = reactance-resistance ratio
f   = frequency in hertz
R   = total resistance in ohms
C   = total capacitance in farads
ω = $2\pi f$

For a capacitor in parallel with a resistor:

$$Q = 2\pi f RC$$

OR

$$Q = \omega RC$$

Q = reactance-resistance ratio
f = frequency in hertz
R = total resistance in ohms
C = total capacitance in farads
ω = $2\pi f$

3.4-11   Capacitive Circuit Steady Current Flow (AC Applications)

$$I = \frac{E_A}{X_C}$$

OR:

$$I = \frac{E_A}{\dfrac{1}{2\pi fC}}$$

I = steady current flow in amperes
E_A = applied voltage in volts
X_C = capacitive reactance in ohms
C = capacitance of applicator

OR: $I = E_A (2\pi fC)$ *ω may be substituted for $2\pi f$ in the formulas

3.4-12   Force Between Two Charges (AC and DC Applications)

$$F = \frac{Q_1 Q_2}{KD^2}$$

F = electrostatic force between plates
Q_1, Q_2 = magnitude of charges
K = dielectric constant
D = distance between plates

3.4-13   Electric Field Intensity (AC and DC Applications)

$$E = \frac{Q}{KD^2}$$

E = electric field intensity
Q = quantity of charge
K = dielectric constant
D = distance between plates

3.4-14   Electric Field Strength (AC and DC Applications)

$$F_s = \frac{V}{D}$$

F_s = field strength
V = potential difference
D = distance moved

3.4-15   Force on Q in Electric Field (AC and DC Applications)

$$F = EQ$$

F = force
E = strength of electric field
Q = magnitude of charge

3.4-16   Potential Energy of Charged Capacitor (AC and DC Applications)

$$PE = \frac{1}{2} CV^2$$

OR

$$PE = \frac{1}{2} \frac{Q^2}{C}$$

PE = potential energy
C = value of capacitance
V = potential difference
Q = quantity of charge

3.4-17   Value of Capacitance (AC and DC Applications)

$$C = \frac{Q}{V}$$

C = value of capacitance
Q = quantity of charge
V = potential difference

3.4-18   Tuning Capacitor's Shunt Capacitance (AC Application Only)

$$Cs = \frac{C_1 F}{2W}$$

Cs = shunt capacitance
C_1 = tuning capacitor's capacitance (change range)
F = low frequency band limit
W = width of frequency band

3.5   CHARACTERISTIC IMPEDANCE (AC APPLICATION ONLY)

A transmission line's primary purpose is to carry electrical energy from its source to its antenna or load with minimum loss. Impedance matching, therefore, between the source and load is most important for maximum power transfer. Since the characteristics of the line may be resistive with either a capacitive or inductive reactive elements when alternating current voltages are applied, the characteristics of a line is generally described by its characteristic impedance (Zo).

3.5-1   Theoritical Value of Purely Resistive Line

$$Z_0 = \sqrt{\frac{L}{C}}$$

OR

$$Z_0 = \frac{Ei}{Ii}$$

Z_0 = characteristic impedance (ohms)
L = inductance per unit length
C = capacitance per unit length
Ei = incident voltage
Ii = incident current

3.5-2   Coaxial Cable Line

Z_0 = characteristic impedance

K = dielectric constant for insulating
   material

*D = inside diameter of outer conductor

*d = outside diameter of inner conductor

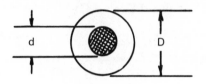

Figure 3-10

*Use the same units for measurements

Solid Dielectric	*Air Dielectric*
$Z_0 = \dfrac{138}{\sqrt{K}} \; \text{Log} \; \dfrac{D}{d}$	$Z_0 = 138 \, \log \, \dfrac{D}{d}$

3.5-3   Coaxial Cable Line Attenuation

$$A = \frac{4.6 \, \sqrt{f} \, (D+d)}{Dd \, (\text{Log} \, \frac{D}{d})} \times 10^{-6}$$

A = attenuation of line (dB/ft)

f = frequency in megahertz

*D = inside diameter of outer conductor

*d = outside diameter of inner conductor

*Same units of measurement

3.5-4   Twin Lead/Parallel Line

Z_0 = characteristic impedance

K   = dielectric constant for insulating
    material

*D = distance between conductor centers

*r = conductor radius

*d = diameter of conductor

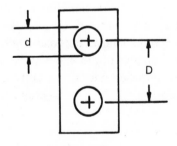

Figure 3-11

*Use same measurements

Solid Dielectric	*Air Dielectric*
$Z_0 = \dfrac{276}{\sqrt{K}} \; \text{Log} \; \dfrac{2D}{d}$	$Z_0 = 276 \, \text{Log} \dfrac{D}{r}$

3.5-5 Standing Wave Ratio

General

$$SWR = \frac{Emax}{Emin}$$

$$SWR = \frac{Imax}{Imin}$$

$$SWR = \sqrt{\frac{Zmax}{Zmin}}$$

SWR = standing wave ratio
E   = voltage on the line
I   = current on the line
Z   = impedance
Z_L = load impedance
Z_o = characteristic impedance

Resistive Load Only

$$SWR = \frac{Z_L}{Z_o} \quad \text{(used if } Z_L \text{ is larger)} \qquad SWR = \frac{Z_o}{Z_L} \quad \text{(used if } Z_o \text{ is larger)}$$

3.5-6 Transformer Matching

$$Z_o = \sqrt{Z_I \, Z_L}$$

Z_o = characteristic impedance
Z_I = imput inpedance
Z_L = output impedance

3.5-7 Reflection Coefficient

$$\Gamma = \frac{Er}{Ei}$$

$$\Gamma = \frac{Ir}{Ii}$$

$$\Gamma = \frac{Z_L - Z_o}{Z_L \, Z_o}$$

Γ = reflection coefficient
Er = reflected voltage
Ei = incident current
Ir = reflected current
Ii = incident current
Z_L = load impedance
Z_o = characteristic impedance

3.5-8 Delay Line Wave Travel T¡me

$$t = N \sqrt{LC}$$

t = the time needed to travel one length of the delay line
N = number of line sections
L = total inductance of one section
C = total capacitance of one section

3.5-9 Pulse Forming Line Pulse Width

$$PW = 2N \sqrt{LC}$$

PW = Pulse width
N   = number of line sections
L   = total inductance of one section
C   = total capacitance of one section

3.5-10   Artificial Line Cutoff Frequency

$$F_{CO} = \frac{1}{\pi \sqrt{LC}}$$

F_{co} = cutoff frequency
π = 3.14
L = total inductance per section
C = total acceptance per section

3.5-11   One-quarter Wavelength Multiple

$$Z_0 = \sqrt{Z_s Z_r}$$

Z_0 = characteristic impedance
Z_s = impedance looking into line source
Z_r = pure resistance load impedance

(line length equals odd multiple of ¼ wavelengths)

3.6   CONDUCTANCE

The property of a component allowing current to flow is termed conductance. The obsolete unit for conductance is mhos (ohm spelled backwards) and was represented by the letter G, while the new metric unit is siemens and is represented by the letter S.

3.6-1   General (DC Applications)

OR
$$G = \frac{1}{R}$$
$$G = \frac{I}{E}$$

G = conductance in mhos
R = resistance in ohms
E = voltage in volts

3.6-2   Resistors in Parallel (DC Applications)

$$G_T = \frac{1}{R_1} + \frac{1}{R_2} + \frac{1}{R_3} + \ldots$$

OR

$$G_T = G_1 + G_2 + G_3 + \ldots$$

3.7   DECIBELS (AC AND DC APPLICATIONS)

Our sense of hearing, based upon an almost logarithmic response, serves as a reference for sound levels. You and I can pretty well distinguish degrees of sound loudness while adjusting the radio, phonograph, tape unit, or TV volume control to a position that meets our individual range of comfortable listening. Since each of our listening levels might differ, a standardized expression denoting relative magnitude of sound level change is desirable. The expression we use is termed *decibel* and is abbreviated officially as dB. However, you may see db, or DB in print.

The decibel provides the means of stating ratios of two voltages, two currents, or two powers.

3.7-1   Currents (Input and Output Impedances Equal)

$$dB = 20 \log \frac{I_2}{I_1}$$

dB = decibel
I_2 = output current
I_1 = input current

3.7-2   Currents (Input and Output Impedances not Equal)

$$dB = 20 \log \frac{I_2 \sqrt{Z_2}}{I_1 \sqrt{Z_1}}$$

dB = decibel
I_2 = output current
I_1 = input current
Z_2 = output impedance
Z_1 = input impedance

3.7-3   Power (Input and Output Impedances Equal)

$$dB = 10 \log \frac{P_2}{P_1}$$

dB = decibel
P_2 = output power
P_1 = input power

3.7-4   Voltages (Input and Output Impedances Equal)

$$dB = 20 \log \frac{E_2}{E_1}$$

dB = decibel
E_2 = output voltage
E_1 = input voltage

3.7-5   Voltages (Input and Output Impedances not Equal)

$$dB = 20 \log \frac{E_2 \sqrt{Z_2}}{E_1 \sqrt{Z_1}}$$

dB = decibel
E_2 = output voltage
E_1 = input voltage
Z_2 = output impedance
Z_1 = input impedance

3.7-6   Reference Levels

The decibel also is used extensively in other areas of electronics which do not rely directly upon voltage, current, or power ratios as did the previous formulas. Whenever the decibel unit is used, make certain you know the reference level employed. If no reference level is given, usually a 6-millivolt signal across a 600-ohm impedance is assumed since it is a standard which corresponds to zero decibel (0 dB). The following are some of the more popular decibel references employed.

Symbol	Reference
dBk	1 kilowatt
dBm	1 milliwatt across 600 ohms
dBV	1 volt
dBW	1 watt
dBVg	voltage gain
dBrap	above Reference Acoustical Power
vu	(volume unit) 1 milliwatt across 600 ohms

3.8   DIRECT CURRENT (AC AND DC APPLICATIONS)

The two voltage potentials encountered daily in electronics are classified as being either AC or DC. The AC potential, previously outlined in sections 3.2-1 through 3.2-17, have current flowing in two alternations whereas DC, being an unidirectional current, theoretically does not change in any appreciable value. Since there are two different classifications for voltages, some of the laws, formulas, and facts applicable to alternating current circuitry are not necessarily suitable for direct current circuitry applications. This section of Chapter 3 will clarify direct current laws, formulas, and facts.

3.8-1   Ohms Laws Variations (DC Application Only)

A. Voltage $E = IR$

$$E = \frac{I}{G} \; or \; E = \frac{I}{S}$$

$$E = \sqrt{PR}$$

$$E = \frac{P}{I}$$

D. Power $P = IE$

$$P = I^2R$$

$$P = \frac{E^2}{R}$$

B. Current $I = \dfrac{E}{R}$

$$I = EG \; or \; I = ES$$

$$I = \sqrt{\frac{P}{R}}$$

C. Resistance $R = \dfrac{E}{I}$

$$R = \frac{P}{I^2}$$

E = voltage in volts
I = current in amperes
R = resistance in ohms
G = conductance in mhos
      (new metric unit of siemens)
P = power in watts

3.8-2   Total Voltage in Resistive Series Circuit (AC and DC Applications)

E_T = total applied voltage (volts)

R   = resistance (ohms)

E_R = potential difference across particular resistor (volts)

$E_T = E_{R1} + E_{R2} + E_{R3} + ...$

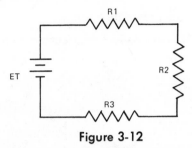

Figure 3-12

3.8-3   Total Voltage in Resistive Parallel Circuit (AC and DC Application)

E_T = total applied voltage (volts)

R = resistance (ohms)

E_R = potential difference across particular resistor (volts)

$E_T = E_{R1} = E_{R2} = E_{R3}$

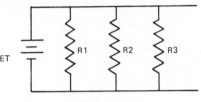

Figure 3-13

3.8-4   Total Current in Series Circuit (AC and DC Application)

I_T = total current (amperes)

I_R = current flowing through that particular resistor (amperes)

R = resistor's resistance (ohms)

$I_T = I_{R1} = I_{R2} = I_{R3}$

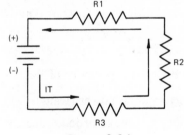

Figure 3-14

3.8-5   Total Current in Parallel Circuit (AC and DC Application)

I_T = total current (amperes)

R = resistors resistance (ohms)

I_R = current flowing through that particular resistor (amperes)

$I_T = I_{R1} I_{R2} + I_{R3} + ...$

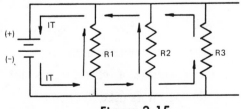

Figure 3-15

3.8-6   Total Resistance in Resistive Series Circuit (AC and DC Application)

R_T= total resistance (ohms)

R_1-R_4 = individual resistances (ohms)

$R_T = R_1 + R_2 + R_3 + R_4 + \ldots$

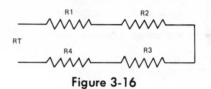

Figure 3-16

3.8-7   Total Resistance in Resistive Parallel Circuits (AC and DC Application)

R_T = total resistance (ohms)

R_1-R_3 = individual resistor's resistance

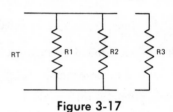

Figure 3-17

A. (For Two) $R_T = \dfrac{R_1 \times R_2}{R_1 + R_2}$

B. (Finding Unknown) $R_2 = \dfrac{R_T \times R_1}{R_1 - R_T}$

C. (For Any Number) $R_T = \dfrac{1}{\dfrac{1}{R_1} + \dfrac{1}{R_2} + \dfrac{1}{R_3} + \ldots}$

3.8-8   Total Power of Resistive Series Circuit (AC and DC Application)

P_T = total power (watts)

P_R = wattage of individual resistor

R = resistor's used in circuit

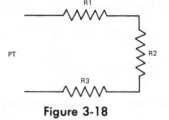

$P_T = P_{R1} + P_{R2} + P_{R3} + \ldots$

Figure 3-18

3.8-9   Total Power of Resistive Parallel Circuit (AC and DC Application)

P_T = total power (watts)

P_R = wattage of individual resistor

R = resistors used in circuit

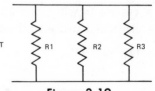

$P_T = P_{R1} + P_{R2} + P_{R3} + \ldots$

Figure 3-19

3.8-10   Total Conductance of Resistive Parallel Circuit (AC and DC Application)

$$G_T = \frac{1}{R_T} \ or \ S = \frac{1}{R_T}$$

$*G_T$ = total conductance (mhos)

OR

R_T = circuitry total resistance

$$G_T = G_1 + G_2 + \ldots$$

$*G_1$ - G_2 = conductance of individual

resistors $\left(\dfrac{1}{R_1} + \dfrac{1}{R_2} \right)$

*Replaced by siemens

3.9   IMPEDANCE (AC APPLICATIONS ONLY)

The laws, formulas, and facts governing current and voltage in direct current circuits seem to be more consistant and less complex than alternating current applications. The main reason for simplicity in DC circuits is because the only opposition offered is that of pure resistance while AC circuit opposition consists of resistance and reactances.

The term describing AC resistance or total opposition to current flow is called impedance. Impedance is mainly due to inherent characteristics of inductors, capacitors, and wire-wound resistors whose oppositions become altered somewhat when alternating voltages or frequencies are applied. You will find data you can rely on when dealing with impedance in this portion of Chapter 3.

3.9-1   Impedance Formulas

(A) Polar Form $\qquad Z\angle\Theta \qquad$ Z = impedance

Θ = direction of angle

(B) Rectangular forms: $\qquad Z = R - jX_C \qquad$ R = resistance

$$Z = R + jX_L \qquad \pm j = \text{``j'' operator}$$

(C) Triangular Forms: $\qquad Z = \sqrt{R^2 + (X_L - X_C)^2} \quad X_C$ = capacitive reactance

X_L = inductive reactance

$$Z = \sqrt{R^2 + (X_C - X_L)^2}$$

(D) Trigonometric Forms: $\qquad (Z) = (\cos\Theta + j\sin\Theta) \quad \cos$ = function of the angle

$$(Z) = (\cos\Theta - j\sin\Theta) \quad \sin = \text{function of the angle}$$

3.9-2   Impedance and Phase Angles for Resistors, Capacitors and Inductors

(A) Resistors:

Figure 3-20

$$Z = R_1$$

$$\Theta = 0°$$

(B) Inductors:

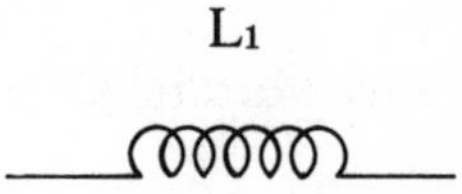

$$L_1$$

Figure 3-21

$$Z = X_L$$

$$Z = 2\pi fL$$

$$\Theta = 90°$$

$$\Theta = +j$$

(C) Capacitors:

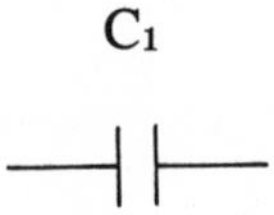

$$C_1$$

Figure 3-22

$$Z = X_C$$

$$Z = \frac{1}{2\pi fC}$$

$$\Theta = -90°$$

$$\Theta = -j$$

3.9-3   Impedance and Phase Angle for Components in Series

(A) Resistors:

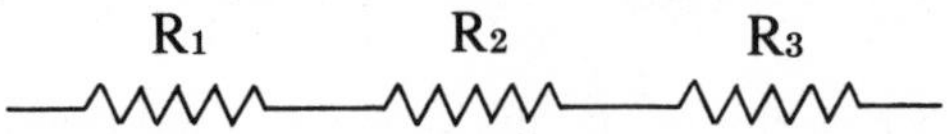

$$R_1 \qquad R_2 \qquad R_3$$

Figure 3-23

$$Z = R_1 + R_2 + R_3 + ...$$

$$\Theta = 0°$$

(B) Inductors:

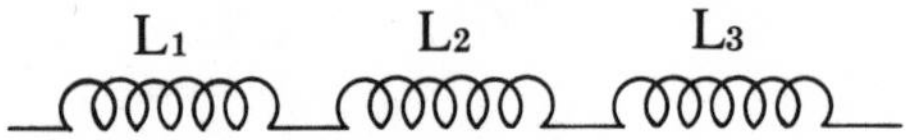

$$L_1 \qquad L_2 \qquad L_3$$

Figure 3-24

$$Z = X_{L1} + X_{L2} + X_{L3} + ...$$

$$\Theta = +90°$$

$$\Theta = +j \text{ (no mutual induction)}$$

(C) Capacitors:

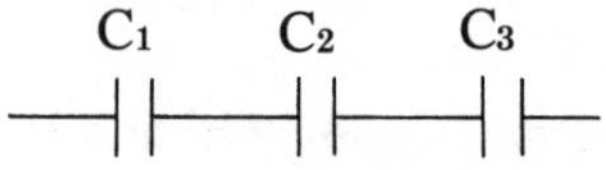

$$C_1 \quad C_2 \quad C_3$$

Figure 3-25

$$Z = X_{C1} + X_{C2} + X_{C3} + ...$$

$$\Theta = -90°$$

$$\Theta = -j$$

3.9-4 Impedance and Phase Angle of Series L-R Circuit

$$Z = \sqrt{R^2 + X_L^2}$$

$$\Theta = \text{arc tan } \frac{X_L}{R}$$

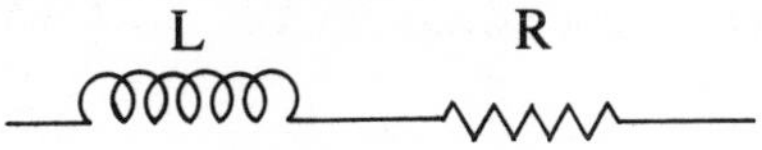

Figure 3-26

3.9-5 Impedance and Phase Angle of Series R-C Circuit

$$Z = \sqrt{R^2 + X_C^2}$$

$$\Theta = \text{arc tan } \frac{X_C}{R}$$

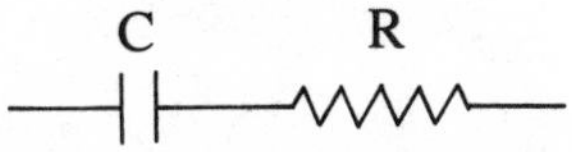

Figure 3-27

3.9-6 Impedance and Phase Angle of Series L-C Circuit

$$Z = X_L - X_C$$
(if X_L is larger)

$$Z = X_C - X_L$$
(if X_C is larger)

$$\Theta = 0° \text{ at resonance}$$
$$(X_L - X_C)$$

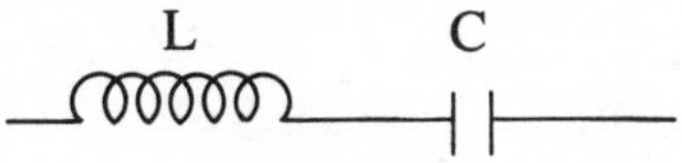

Figure 3-28

3.9-7 Impedance and Phase Angle of Series R-L-C Circuit

$$Z = \sqrt{R^2 + (X_L - X_C)^2}$$

$$Z = \sqrt{R^2 + (X_C - X_L)^2}$$

$$\Theta = \text{arc tan } \frac{X_L - X_C}{R}$$

Figure 3-29

$$\Theta = \text{arc tan } \frac{X_C - X_L}{R}$$

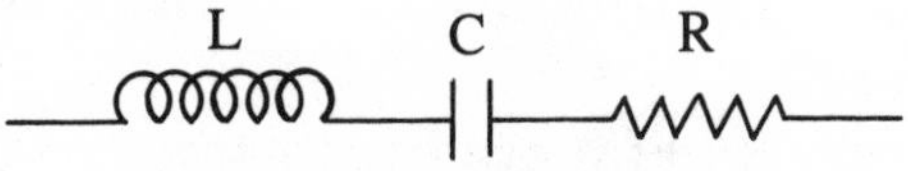

3.9-8 Reactive Circuit Summary

(A) *R-L Circuit:* (Voltage leads current)

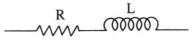

Figure 3-30

$$X_L = \omega L = 2 \pi f L$$

$$Z = \sqrt{R^2 + X_L^2}$$

$$I_T = \frac{E \text{ applied}}{Z}$$

X_L = inductive reactance (ohms)
$\omega = 2 \pi f$
R = pure resistance (ohms)
E applied = source potential (volts)
Z = circuitry impedance (ohms)
I_T = total current (amperes)

(B) *RC Circuit:*
    (Current leads voltage)

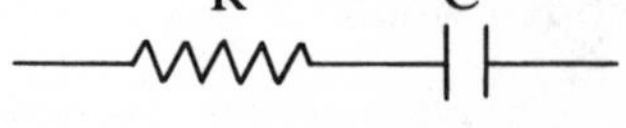

Figure 3-31

$$X_C = \frac{1}{\omega C} = \frac{1}{2\pi fC} = \frac{0.159}{fC}$$

$$Z = \sqrt{R^2 + X_C^2}$$

$$I_T = \frac{E \text{ applied}}{Z}$$

X_C = capacitive reactance (ohms)
$\omega = 2\pi f$
R = pure resistance (ohms)
E applied = source potential (volts)
Z = circuitry impedance (ohms)
I_T = total current (amperes)

(C) *R-L-C Circuit*

Figure 3-32

$$X_C = \frac{1}{\omega C} = \frac{1}{2\pi fC} = \frac{0.159}{fC}$$

$$X_L = \omega L = 2\pi fL$$

$$Z = \sqrt{R^2 + X_T^2}$$

$$*Z = \sqrt{R^2 + (X_L - X_C)^2}$$

$$*Z = \sqrt{R^2 + (X_C - X_L)^2}$$

$$I_T = \frac{E, \text{applied}}{Z}$$

X_C = capacitive reactance (ohms)
$\omega = 2\pi f$
X_L = inductive reactance (ohms)
L = value of inductor (henrys)
C = value of capacitor (farads)
R = pure resistance (ohms)
X_T = total reactance (ohms)
I_T = total current (amperes)
Z = impedance (ohms)
E applied = source potential (volts)

*subtract smaller reactance from larger
(circuit acts inductive or capacitive depending upon
largest opposition)

3.9-9   R-L-C Resonant Series Circuitry Data

$$F_r = \frac{1}{2\pi\sqrt{LC}} = \frac{0.159}{\sqrt{LC}}$$

$$Z = R$$

$$X_L = X_C$$

$$\Theta = 0°$$

F_r = resonant frequency (hertz)
L = value of inductor (henrys)
C = value of capacitor (farads)
Z = impedance (ohms)
R = pure resistance (ohms)
X_L = inductive reactance (ohms)
X_C = capacitive reactance (ohms)

(current is maximum, circuit acts
resistive)

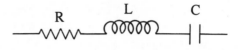

Figure 3-33

     The characteristics of the series R-L-C circuit illustrated in Figure 3-33 will change somewhat when the frequency goes above or below the particular resonant frequency. The following chart summarizes how the circuitry acts at frequencies above, below and at the desired frequency.

	Above Resonance	At Resonance	Below Resonance
The circuit appears to be:	X_L + R	Resistive	X_C + R
The circuitry current is:	Low	High	Low
The circuitry impedance is:	High	Low	High
The circuitry current phase is:	Lagging E	in phase	Leading E

3.9-10   Impedance and Phase Angle for Components in Parallel

(A) *Resistors:*

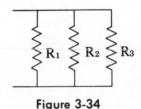

Figure 3-34

$$Z = \frac{R_1 \times R_2}{R_1 + R_2} \text{ (when only two resistors)} \qquad Z = \frac{1}{\dfrac{1}{R_1} + \dfrac{1}{R_2} + \dfrac{1}{R_3}} \text{ (more than two resistors)}$$

$$\Theta = 0° \qquad\qquad\qquad\qquad\qquad \Theta = 0°$$

(B) *Inductors:*

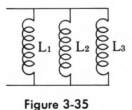

(no mutual induction)   Figure 3-35

$$Z = \frac{(X_{L1})(X_{L2})}{X_{L1} + X_{L2}} \text{ (when only two inductors)}$$

$$Z = \frac{1}{\dfrac{1}{L_1} + \dfrac{1}{L_2} + \dfrac{1}{L_3} + \ldots}$$ (when more than two inductors)

$\Theta = +90°$

$\Theta = + j$

(C) *Capacitors:*

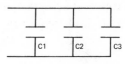

$$Z = \frac{(X_{C1})(X_{C2})}{X_{C1} + X_{C2}}$$ (when only two capacitors)

Figure 3-36

$$Z = \frac{1}{\dfrac{1}{X_{C1}} + \dfrac{1}{X_{C2}} + \dfrac{1}{X_{C3}} + \ldots}$$ (when more than two capacitors)

$\Theta = -90°$

$\Theta = - j$

3.9-11   Impedance and Phase Angle of Parallel R-L Circuit

$$Z = \frac{(R)(X_L)}{\sqrt{R^2 + X_L^2}}$$

$$\Theta = \text{arc tan} \frac{R}{X_L}$$

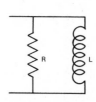

Figure 3-37

3.9-12   Impedance and Phase Angle of Parallel R-C Circuit

$$Z = \frac{(R)(X_C)}{\sqrt{R^2 + X_C^2}}$$

$$\Theta = \text{arc tan} \frac{R}{X_C}$$

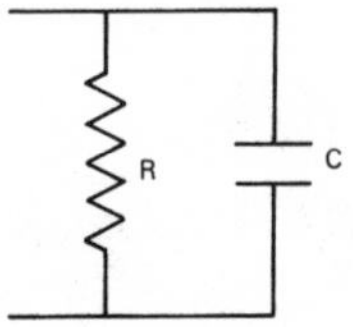

Figure 3-38

3.9-13   Impedance and Phase Angle of Parallel L-C Circuit

$$Z = \frac{(X_L)(X_C)}{X_L - X_C}$$

$$Z = \frac{(X_C)(X_L)}{X_C - X_L}$$

$\Theta = 0°$ at resonance ($X_L = X_C$)

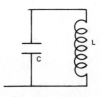

Figure 3-39

3.9-14   Impedance and Phase Angle of Parallel L-C-R Circuit

$$Z = \frac{(R)(X_L)(X_C)}{\sqrt{X_L^2 \, X_C^2 + R^2 \, (X_C - X_C)^2}}$$

$$\Theta = \frac{R(X_L - X_C)}{(X_L)(X_C)}$$

*The formulas assumed X_L to the larger than X_C. Change values if X_C is larger.

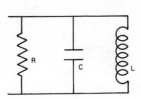

Figure 3-40

3.9-15   Parallel Reactive Circuit Summary

(A) *R-L Circuit:* (current lags voltage)

$$X_L = 2\,\pi\,fL = \omega L$$

$$.Z = \frac{E \text{ applied}}{I_{Total}}$$

$$Z = \frac{(R)(X)}{\sqrt{R^2 + X^2}}$$

$$I_T \sqrt{I_R^2 + I_{XL}^2}$$

X_L = inductive reactance (ohms)
$\omega = 2\,\pi\,f$
f = frequency (hertz)
L = value of indicator (henrys)
Z = impedance (ohms)
E applied = source potential volts
I_{Total} = total circuitry current (amperes)
R = pure resistance (ohms)
X = Total reactance (ohms)

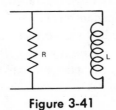

Figure 3-41

(B) *R-C Circuit:*
(voltage lags current)

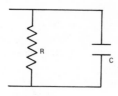

Figure 3-42

$$X_C = \frac{1}{2 \pi f C} = \frac{1}{\omega C} = \frac{0.159}{f C}$$

$$Z = \frac{E\ applied}{I_{Total}}$$

$$I_T = \sqrt{I_R^2 + I_{XC}^2}$$

X_C = capacitive reactance (ohms)
$\omega = 2 \pi f$
E applied = source potential (volts)
I_{Total} = total circuitry current (amperes)
R = pure resistance (ohms)
Z = impedance (ohms)

(C) *R-L-C Circuit:*

*circuit acts inductive or capacitive depending upon which leg has largest current.

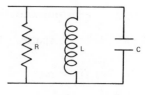

Figure 3-43

$$X_C = \frac{1}{2 \pi f C} = \frac{1}{\omega c} \quad \frac{0.159}{f C}$$

$$X_L = 2 \pi f L = \omega L$$

$$Z = \frac{E\ applied}{I_{Total}}$$

$$I_T = \sqrt{I_R^2 + I_X^2}$$

$$I_T = \sqrt{I_R^2 + (I_{XL} - I_{XC})^2}$$

$$I_T = \sqrt{I_R^2 + (I_{XC} - I_{XL})^2}$$

X_C = capacitive reactance (ohms)
$\omega = 2 \pi f$
X_L = inductive reactance (ohms)
Z = impedance (ohms)
E applied = source potential (volts)
I_{Total} = Total circuitry current (amperes)
X = total reactance (ohms)

3.9-16   R-L-C Resonant Parallel Circuitry Data

(current is minimum, circuit acts
resistive)

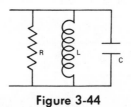

Figure 3-44

$$Fr = \frac{1}{2\pi\sqrt{LC}} = \frac{0.159}{\sqrt{LC}}$$

$$Z = R$$

$$X_C = X_L$$

$$I_L = I_C$$

$$\Theta = 0°$$

Fr = resonant frequency (hertz)
L = value of inductor (henrys)
C = value of capacitor (farads)
Z = impedance (ohms)
R = pure resistance (ohms)
I = current (amperes)
X_C = capacitive reactance (ohms)
X_L = inductive reactance (ohms)

The following chart further summarizes the inherent characteristics the circuit of Figure 3-44 has when the frequency of the circuit goes above and below the desired frequency.

	Above Resonance	At Resonance	Below Resonance
The circuit appears to be:	$X_C + R$	Resistive	$X_L + R$
The circuitry current is:	High	Low	High
The circuitry impedance is:	Low	High	Low
The circuitry current phase is:	Leads E	in phase	lags E

3.9-17   Impedance Matching

(A) *High input to Lower Value Output Resistance*

when $R_{in} > R_0$

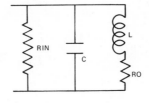

Figure 3-45

$$X_L = \sqrt{R_0 R_{in} - R_0^2}$$

$$X_C = \frac{R_0 R_{in}}{X_L}$$

X_L = inductive reactance (ohms)
R_0 = output or load resistance (ohms)
R_in = input resistance (ohms)
X_C = capacitive reactance (ohms)

(B) *Low input to High Value Output Resistance*

when $R_{in} < R_0$

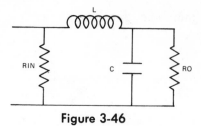

Figure 3-46

$$X_C = R_L \sqrt{\frac{R_{in}}{R_0 - R_{in}}}$$

$$X_L = \frac{R_0 R_{in}}{X_C}$$

R_{in} = input resistance (ohms)
R_0 = output or load resistance (ohms)
X_C = capacitive reactance (ohms)
X_L = inductive reactance (ohms)

3.10   INDUCTORS (AC AND DC APPLICATIONS)

Transformers, chokes, toroidal coils and ferrite beads all have one thing in common; and that is the family type of grouping called inductors. In its simplest form, an inductor or choke or coil, as it is sometimes called, relies upon magnetic fields produced when alternating current flows through them. The ability to an inductor to oppose current changes causes a smoothing out action to occur.

The unit of inductance is the Henry, but smaller units of millihenry and microhenry are used extensively in electronic application. Inductance itself is dependent upon the number of turns, cross-sectional dimensions of the coil, length of the winding, and permeability of the coil core. All of the following formulas and facts will aid you when working with inductors.

3.10-1   Inductors in Series

(A) *Total Inductance-no mutual inductance*

(AC and DC application)

Figure 3-47

$$L_T = L_1 + L_2 + L_3 + \ldots$$

L_T = total inductance (henrys)
$L_1 - L_3$ = individual inductor values (henrys)

(B) *Total Inductance with Mutual or Coupled Inductance*

(AC application only)

Figure 3-48

$L_T = L_1 + L_2 + 2m$ L_T = total inductance (henrys)

$L_T = L_1 + L_2 - 2m$ $L_1 - L_2$ = individual coil inductance (henrys)

+ 2m = mutual inductance, fields aiding

- 2m = mutual inductance, fields opposing

$$m = \frac{L_A - L_B}{4}$$

L_A = total inductance of L_1, L_2 with fields opposing

L_B = total inductance of L_1, L_2 with fields opposing

3.10-2   Inductors in Parallel

(A) *Total Inductance of Two Inductors (No mutual inductance)*

(AC and DC application)

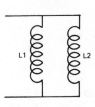

Figure 3-49

$$L_T = \frac{L_1 L_2}{L_1 + L_2}$$ L_T = total inductance (henrys)

$L_1 - L_2$ = individual coil inductance (henrys)

(B) *Total Inductance of Two Inductors with Mutual Inductance*

(AC application only)

Figure 3-50

$$L_{TA} = \cfrac{1}{\cfrac{1}{L_1 + m} + \cfrac{1}{L_2 + m}}$$

$$L_{TB} = \cfrac{1}{\cfrac{1}{L_1 - m} + \cfrac{1}{L_2 - m}}$$

L_{TA} = Total inductance with fields aiding

L_{TB} = Total inductance with fields opposing

$L_1 - L_2$ = individual coil inductance (henrys)

m = mutual inductance

(C) *Total Inductance of any number of Inductors (no mutual inductance)*

(AC and DC application)

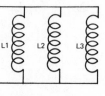

$$L_T = \cfrac{1}{\dfrac{1}{L_1} + \dfrac{1}{L_2} + \dfrac{1}{L_3} + \, ...}$$

Figure 3-51

L_T = total inductance (henrys)
L_1-L_3 = individual coil inductance (henrys)

(D) *Total Inductance of Two Separated Coils with Mutual Inductance*

(AC application only)

Figure 3-52

$$L_T = L_1 + L_2 \pm 2m$$

L_T = total inductance (henrys)
L_1-L_2 = individual coil inductance (henrys)
m = mutual inductance

3.10-3   Mutual Inductance (AC Application Only)

$$m = \frac{L_A - L_B}{4}$$

$$m = K \sqrt{L_A L_B}$$

L_A = total inductance with fields aiding
L_B = total inductance with fields opposing
K = coefficient of coupling

3.10-4   Coefficent of Coupling (AC applications only)

(A) $\quad K = \dfrac{m}{\sqrt{L_A L_B}}$

(B) $\quad K = \dfrac{m}{\sqrt{L_p L_s}}$

K = coupling coefficient
L_A, L_B = individual coil inductance (henrys)
m = mutual inductance
L_p = primary's coil inductance (henrys)
L_s = secondary's coil inductance (henrys)

3.10-5   Inductance (AC and DC Applications)

(A) *General*

$$L = \frac{N}{I} \times 10^{-8}$$

L = inductance (henrys)
N = number of turns
I = current in amperes

(B) *Single Layer-Air Core*

$$LT = \frac{R^2 N^2}{9R + 19(l)}$$

LT = inductance in microhenrys
R = radius of coil (inches)
N = number of turns
l = length of coil (inches)

3.10-6   Number of Turns Required (AC and DC Applications)

$$N = \sqrt{\frac{L[9R + 10](l)}{R^2}}$$

N = number of turns
L = inductance in microhenrys
R = radius of coil in inches
l = length of coil in inches

3.10-7   Number of Feet Required (AC and DC Applications)

$$F = \frac{RD}{Rg} (L)$$

F = number of feet
RD = desired resistance
Rg = resistance per given length
            (1000 feet normally)
L = length used to determine resistance

3.10-8   Energy Stored in an Inductor (AC and DC Applications)

$$W = \frac{LI^2}{2}$$

W = energy in joules
L = inductance in henrys
I = steady current in amperes

3.10-9   Average Induced Voltage (AC Application Only)

(A) into a coil

$$eAV = \frac{-N \Delta \Theta}{\Delta t} \times 10^{-8}$$

eAV = average induced voltage
N = number of turns in coil
$\Delta\Theta$ = change in flux (maxwell)
Δt = time increment given for flux change
            (seconds)

(B) *From one coil into another*

$$eAV = -M\frac{\Delta i}{\Delta t}$$

eAV = average induced voltage
M   = coefficient of mutual inductance
        between coils
Δi = change in current (amperes)
Δt = time increment (seconds)

Note: The minus sign indicates that the induced voltage opposes the force that created it and may or may not be included in the equations.

3.10-10   Counter EMF Across an Inductor (AC Application Only)

$$eL = L\frac{\Delta i}{\Delta t}$$

eL   = instantaneous voltage (volts)
L   = value of inductance (henrys)
Δi = change in current (amperes)
Δt = change in time (seconds)

3.11   MAGNETISM—ELECTROMAGNETS (AC APPLICATION ONLY)

A phenomenon often desirable in electronic circuitry is the generation of magnetic fields created when wire conductors and coils have current flow. The magnetism and electromagnetism terms encountered include magnetic poles, fields, force, flux, permeability, and hysteresis. Some of the formulas relating to these terms are given within this section of Chapter 3.

3.11-1   Magnetic Field Intensity

(A) *General*

$$H = \frac{F}{m}$$

*H   = field strength in oersteds
F   = force exerted on pole by the field
m   = strength of pole

(B) *Around Straight Conductor*

$$H = \frac{2I}{10R}$$

*H   = field strength (oersteds)
I   = current flowing (amperes)
R   = radius of conductor (centimeters)

(C) *Long Coils*

$$H = \frac{4\pi NI}{10L}$$

*H   = field strength (oersteds)
N   = number of turns
I   = current flowing (amperes)
L   = length of coil (centimeters)

(D) *Short Coils*

$$H = \frac{2\pi NI}{10R}$$

*H = field strength (oersteds)
N = number of turns
I = current flowing (amperes)
R = radius of coil (centimeters)

*Oersted unit is superseded by the metric unit ampere/metre.

3.11-2   Force Between Two Magnetic Poles

$$F = \frac{M_1 M_2}{\mu d^2}$$

F   = force
M_1 = strength of one pole
M_2 = strength of second pole
μ = permeability of medium
d   = distance between the poles

3.11-3   Permeability

$$\mu = \frac{\beta}{H}$$

*μ = permeability
β = flux density
H = field density

*indicates ease with which magnetic lines of force flow in a magnetic circuit.

3.11-4   Flux

$$\emptyset = \frac{mmf}{R}$$

$\emptyset$ = flux in maxwells (webers)
mmf = magnetomotive force in gilberts (ampere turns)
R   = reluctance

3.11-5   Magnetomotive Force

$$mmf = \frac{4\pi NI}{10}$$

mmf = magnetomotive force
N   = number of turns
I   = current flowing

3.11-6   Reluctance

$$R = \frac{L}{\mu A}$$

R = reluctance
μ = permeability
A = cross sectional area in square centimeters
L = coil length in centimeters

3.11-7 Flux Density

$$\beta = \frac{\emptyset}{A}$$

β = flux density in gauss (Tesla)
$\emptyset$ = flux in maxwells (Webers)
A = cross sectional area in square centimeters

3.11-8 Force Exerted on a Conductor

$$F = \frac{\beta\, IL}{10} \sin \Theta$$

F = force in dynes (Newtons)
β = field strength in gauss (Tesla)
I = conductor current in amperes
L = length of conductor in sq. centimeters
Θ = conductor's angle with the flux-field strength

3.12   METERS

An ammeter and voltmeter differ mainly in the circuitry design and component positioning within the meter's enclosure. If the meter serves as an ammeter, there will be a resistor or resistors in parallel (shunt) with the meter, while a voltmeter uses a resistor or resistors connected in series with the meter. Since series circuitry forms a voltage divider and parallel circuitry forms a current divider, only additional knowledge of ohms law needs to be applied when altering meter ranges. The useful data, formulas, and illustrations found in this section of Chapter 3 will help you design, troubleshoot, or modify your meter when the need arises.

3.12-1 Meter Sensitivity

Sensitivity, when speaking about meter movements, is a term used to describe the current-voltage required for full-scale deflection. Sensitivity of a voltmeter is the ability a meter movement has which allows a small amount of voltage to be measured without loading or in any appreciable way disturbing the circuitry under test. This sensitivity is measured in units of ohms-per-volt; the amount of resistance required for a one-volt meter movement.

Voltmeters having good sensitivity begin around 20,000 ohms-per-volt. The following formulas can be used when determining meter sensitivity.

(A) *Voltmeter*

$$\frac{R}{E} = \Omega/V = \frac{R_m}{V_{Fs}}$$

$\dfrac{R}{E}$ or $\dfrac{\Omega}{V}$ = ohms-per-volt

(B) *Ammeter*

$$\frac{\Omega}{V} = \frac{1}{I_{Fs}}$$

R_m = meter resistance
V_{Fs} = fullscale reading voltage
I_{Fs} = fullscale current reading

3.12-2   Determining Meter Resistance

You can determine DC meter resistances using two variable resistors, a dc power source, and an ohmmeter if you carefully apply the following hints. Construct the circuit illustrated in Figure 3-53, making certain the switch is open and R_1 is set for maximum resistance.

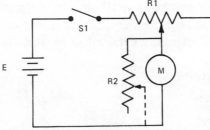

Figure 3-53

After making certain R_1 is set at maximum, close switch S_1 and adjust R_1 until the meter indicates full-scale reading; then connect resistor R_2 in parallel with with meter, adjusting it until the meter deflection ins one-half its original full-scale reading. Open switch S_1, disconnect one side of R_2 from the circuit and then measure the resistance of R_2. The value obtained represents the meters resistance.

3.12-3   Ammeters

Ammeter ranges can be extended by shunting the meter with a particular value resistor. To calculate the value for a shunt resistor, we must know the maximum current value to be measured and the maximum current required for full-scale deflection before applying Kirchoff's current law. These concepts are as follows:

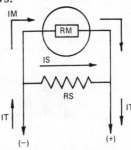

$$Rs = \frac{Rm}{N-1}$$

$$Rs = \frac{ImRm}{IT-Im}$$

$$Rs = \frac{ImRm}{Is}$$

Figure 3-54

Rs = shunt resistor (ohms)

IT = total current (amperes)

Is = shunt resistor's current (amperes)

Im = meter's fullscale current (amperes)

Rm = meter's resistance (ohms)

*N = scale multiplication factor

*This is the new full scale reading divided by the original full scale reading (both in same units).

3.12-4   Voltmeters

The series resistive circuitry commonly encountered within a voltmeter forms a voltage divider circuit. These resistors, called multipliers, are added to the meter to increase or extend the meter's voltage range. Since voltmeters are connected in parallel with voltage potentials and because all circuitry currents flow through the multiplier resistors and then on through the meter, calculations needed when determining values for multiplier resistors are easily obtained using ohms law.

R_1, R_2 = multiplier resistors

R_m = meter resistance

E_{fs} = full scale voltage

I_{fs} = full scale current

For example sake:

R_m = 10 k Ω

E_{fs} = 10 volts

I_{fs} = 0.001 amperes

R_x = resistor for position #1

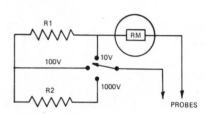

Figure 3-55

(A) *(General Formula) Position #1, 10 Volts*

$$R_X = \frac{E_{FS}}{I_{FS}} - R_m$$

$$R_X = \frac{10v}{1mA} - 10k\ \Omega = 10k\ \Omega - 10k\ \Omega$$

$R_X = 0$ ohms (none required because E_{fs} = 10 volts)

(B) *1st General Formula, Position #2, 100 volts*

$$R_1 = \frac{E_{fs}}{I_{fs}} - R_m = \frac{100v}{1mA} - 10k\ \Omega = 100k\ \Omega - 10k\ \Omega$$

$R_1 = 90k\ \Omega$

2nd General Formula, Position #2, 100 volts

$$R_1 = R_m (N-1) \text{ where } N = \frac{\text{new voltage}}{\text{old voltage}}$$

$$R_1 = 10k\ \Omega \frac{(100-1)}{10} = 10k\ \Omega\ (9)$$

$R_1 = 90k\ \Omega$

(C) *1st General Formula, Position #3, 1000 volts*

$$R_2 = \frac{E_{fs}}{I_{fs}} - R_m + R_1 = \frac{1000v}{1\ mA} - 10k\ \Omega + 90k\ \Omega$$

$$R_2 = 1M - 100k\ \Omega = 900k\ \Omega$$

2nd General Formula, Position #3, 1000 volts

$$R_2 = R_m + R_1\ (N-1)\ \text{where}\ N = \frac{\text{new voltage}}{\text{old voltage}}$$

$$R_2 = 10k\ \Omega + 90k\ \Omega\frac{(1000 - 1)}{100} = 100k\ \Omega\ (9)$$

$$R_2 = 9000k\ \Omega$$

It should be noted from Figure 3-55 and from the formulas for each of the three positions denoting a voltage range increase from 10 volts to 100 volts and finally to 1000 volts, that the basic formula was modified somewhat for each range. You will need to add additional values of resistances to the basic formulas since a voltage divider action exists. To sum up this on voltmeters, the third position formula would be $\frac{E_{fs}}{I_{fs}} - R_m + R_1 + R_2$ or $R_m + R_1 + R_2\ (N-1)$.

3.13   MODULATION (AC APPLICATIONS ONLY)

The data found in this section of Chapter 3 will highlight amplitude and frequency modulation. Since the terminology for modulation is similar, they will be pointed out prior to AM or FM formula particulars.

Modulation, whether it is amplitude, frequency, or phase, generates a new set or sets of radio frequencies which are symmetrically distributed around a carrier frequency.

When audio frequencies are used to control the amplitude of a carrier radio frequency, the general term used to describe this process is *amplitude modulation.*

Amplitude modulation creates frequencies which are actually sums and differences of the frequencies used. These two particulars, when generated, are called *beat frequencies.* The frequency sum is termed *upper side/side band frequency* and the difference is the *lower side/side band frequency.*

When the instantaneous values of the separate frequencies (lower and upper) are added together in modulation, a *modulation envelope* is created. The AM envelope resembles the amplitude variations of the signal used when modulating the carrier frequency. FM does not actually use the term envelope because after modulation there is no amplitude change, only a frequency change within the carrier's frequency amplitude.

The reference for frequency modulation is the unmodulated transmitter oscillator's frequency termed *center frequency.* When the modulating signal causes the frequency of the modulated wave to increase or decrease above or below the center frequency, the effect is called *frequency deviation or swing. Bandwidth* is the total frequency range which exists between minimal and maximum frequency deviations.

3.13-1   Modulation Factor (AM and FM)

(A) *Amplitude Modulation (General)*

$$M = \frac{Es}{Ec} \times 100$$

M = modulation factor (%)
Es = modulating signal amplitude
Ec = carrier signal amplitude

(100% modulation or less)

$$M = \frac{Emax - Emin}{Emax + Emin} \times 100$$

$$M = \frac{Emax - Emin}{2Eav} \times 100$$

M = modulation factor (%)
Emax = maximum amplitude of carrier
Emin = minimum amplitude of carrier
Eav = average amplitude

(more than 100% modulation)

$$M = \frac{Emax - Ec}{Ec} \times 100$$

M = modulation factor (%)
Emax = maximum amplitude
Ec = carrier amplitude

(B) *Frequency Modulation*

$$M = \frac{\Delta F}{\Delta F@100\% m} \times 100$$

M = modulation factor (%)
ΔF = frequency deviation

3.13-2   Modulation Index (FM Only)

$$MI = \frac{Fd}{Fm}$$

$$\beta = \frac{\Delta f}{Fm}$$

MI, β = modulation Index
Fd = frequency deviation
Fm = modulating frequency
Δf = frequency shift from unmodulated carrier

3.13-3   Bandwidth (AM and FM)

(A) *Amplitude Modulation*

$$BW = 2FM$$

BW = bandwidth
FM = highest modulating frequency
β = modulation index

(B) *Frequency Modulation*

$$BW = 2FM (1 + \beta)$$

Note: AM has only one pair of sidebands, FM has several pairs. The modulation index indicates the number of FM sideband pairs; the higher the index, the more sideband pairs there are.

3.13-4   Power (AM Only)

(A) *Sideband Power*

$$Psb = \frac{M^2}{2} \times Pc$$

Psb = sideband power (both sidebands)
M = percent of modulation
Pc = carrier power

(B) *Total Radiated Power*

$$PT = Psb + Pc$$

PT = total power
Psb = sideband power (both sidebands)
Pc = carrier power

3.14   OSCILLATORS (AC APPLICATION ONLY)

If one were to group the many existing oscillators into some meaningful classification, four specific groupings would probably evolve. One grouping would be classified according to circuitry, another would be based upon the output waveform obtained, the third would be according to the frequency of the generated wave, and the fourth might be according to the method used to excite the oscillator circuitry. One method, called impulse or shock excitation, produces oscillations when the duration of applied voltage is short in comparison with the duration of current produced. It should be pointed out that sharp contrasting differences defining each group would be very difficult to make because there are so many with similar characteristics.

Amplifiers and oscillators are also similar since both require the amplification qualities offered by vacuum tubes or transistors. In order to have an oscillator, an amplifier will have a feedback loop through which positive voltages or currents are coupled back to the amplifier's input circuitry. The formulas relating to oscillators are as follows:

3.14-1   L-C Oscillator

$$F_0 = \frac{1}{2\pi\sqrt{LC}}$$

$$F_0 = \frac{0.159}{\sqrt{LC}}$$

F_0 = output frequency (Hertz)
L = value of inductor (henrys)
C = value of capacitance (Farads)

3.14-2   Phase Shift Oscillator

$$F_0 = \frac{1}{2\pi RC\sqrt{6}}$$

F_0 = output frequency (Hertz)
R = value of resistances (ohms)
C = value of capacitors (Farads)

3.14-3   Wien Bridge Oscillator

$$F_0 = \frac{1}{2\pi\sqrt{R_1 C_1 R_2 C_2}}$$

$$*F_0 = \frac{1}{2\pi R_1 C_1}$$

F_0 = output frequency (Hertz)
R_1, R_2 = resistances in arms (Ohms)
C_1, C_2 = capacitance in arms (Farads)

*use when $R_1 = R_2$ and $C_1 = C_2$

3.14-4   Shock-Excited Oscillator Specifics

(A) *Merit of Oscillator Tank Circuit*

$$Q = \frac{R_S}{X_L} = \frac{R_S}{X_C}$$

$$Q = \frac{R_S}{L/C}$$

Note: At resonance
$X_L = X_C = L/C$

Q = merit
R_S = shunt resistance (ohms)
X_L = inductive reactance (ohms)
X_C = capacitive reactance (ohms)
L = value of inductor (henrys)
C = value of capacitor (Farads)

(B) *Output Frequency when Q equals 5 or Greater*

$$F_0 = \frac{1}{2\pi\sqrt{LC}}$$

F_0 = output frequency (Hertz)
L = value of inductor (henrys)
C = value of capacitor (Farads)

(C) *Output Frequency When Q is less than 5*

$$F_0 = \frac{1}{2\pi\sqrt{LC}} \sqrt{1 - \left(\frac{1}{2Q}\right)^2}$$

F_0 = output frequency (Hertz)
L = value of inductor (henrys)
C = value of capacitor (Farads)
Q = merit

3.15   POWER (AC AND DC APPLICATIONS)

An important circuitry design factor used extensively in electronics is power. Some of the formulas used in DC applications do not apply to AC circuitry due to the complex nature of alternating current and circuit component characteristics. The data in this section of Chapter 3 will provide unmistakable power results when selecting the right formula.

3.15-1   Electrical Work

$$W = P \times T$$

W = energy in watt-hours
P = power in watts
T = time in hours

3.15-2 DC Formulas

$P = IE$	P = power in watts
$P = I^2R$	I = current in amperes
$P = \dfrac{E^2}{R}$	E = voltage in volts
	R = resistance in ohms

3.15-3 AC Formulas

(A) *Apparent Power*

$P = IE$	P = effective power in watts
$P = I^2R$	I = effective current in amperes
$P = \dfrac{E^2}{R}$	E = effective voltage in volts
	R = resistance or impedance in ohms

(B) *True Power*

$P = \dfrac{E^2 \text{Cos}\Theta}{Z}$	P = true power in watts
	E = voltage in volts
$P = IE \text{ Cos } \Theta$	$\text{Cos } \Theta$ = power factor
$P = I^2Z \text{ Cos } \Theta$	Z = impedance in ohms
$P = PaPf$	I = current in amperes
	Pa = apparent power in watts
	Pf = power factor

(C) *Current*

$I = \dfrac{P}{E \cos \Theta}$	I = current in amperes
	P = power in watts
	E = voltage in volts
$I = \sqrt{\dfrac{P}{Z \cos \Theta}}$	Z = impedance in ohms
	$\text{Cos } \Theta$ = power factor

(D) *Impedance*

$Z = \dfrac{E^2 \cos \Theta}{P}$	Z = impedance in ohms
	E = voltage in volts
	$\cos \Theta$ = power factor
	P = power in watts

(E) *Voltage*

$E = \sqrt{\dfrac{PZ}{\cos \Theta}}$	E = voltage in volts
	P = power in watts
	Z = impedance in ohms
	$\cos \Theta$ = power factor

3.16   POWER SUPPLIES

Power sources or power supply circuitry, being the heart of all electronic devices, tend to be the first place an electronic professional checks when particular symptoms arise. This is rightly so because without voltage applied (AC or DC) to active components, the electronic device is useless. The following facts, figures, and formulas will provide quick access to data necessary for power supply circuitry design or troubleshooting.

3.16-1   Terminology

(A) *Bleeders:* This term describes a resistor whose purpose is to discharge capacitors in the power supply circuitry, thereby avoiding physical shocks when power is turned off. The resistor value is large, so small amounts of circuitry current are consumed. Normally, its value is based upon a current value of 10 percent or less of the total output current value.

(B) *Input Resistance*

This term is used when analyzing transformer impedances combined with rectifier resistances. Sometimes it is called input impedance or input resistance. The generalized formulas used, providing close approximations for half-wave and full-wave transformer type power supplies, are:

HALF WAVE

$$R_{in} = N^2R_1 + R_2$$

R_{in} = input impedance/resistance (transformer impedance plus rectifier resistance)
R_1 = primary resistance
R_2 = secondary resistance
N = primary to secondary turns ratio

FULL WAVE

$$R_{in} = N^2R_1 + R_2$$

R_{in} = input impedance/resistance
R_1 = primary resistance
R_2 = one-half secondary total resistance
N = primary to one-half secondary turns ratio

(C) *Load or Output Resistance*

This term, when talking about power supplies, defines the over-all effect felt at the filter circuit, or specifically at the power supply's output terminals. Since circuitry connected to the output terminals requires so much current at so much voltage, the output resistance is easily calculated using ohms law.

$$RL = \frac{Eo}{IT}$$

RL = load or output resistance
Eo = output voltage
IT = total full load current

(D) *Peak Inverse Voltage*

The voltage at which diodes will break down when not conducting is called peak inverse voltage, peak reverse voltage, breakdown voltage or something similar. It is the maximum voltage present or allowable on the anodes and cathodes of diodes when a reversed biased condition normally results. When buying diodes, note that one of the ratings will indicate this maximum potential before destruction.

Peak inverse voltage limits may be extended when diodes are placed in series, but, their current limits must be adequate for circuitry demands. When diodes are connected in parallel, the peak inverse voltage is not extended, but additional amounts of current are obtainable.

(E) *Ripple*

To the untrained, it might seem that all electronic devices which plug into an alternating source operate on alternating current, but this is not always true. Electronic circuitry normally demands some amount of direct current potential to operate as designed; therefore, AC to DC rectification is necessary.

Ripple relates to the pulsations of AC which are superimposed on steady DC potentials. The measurement of power supply ripple tells the effectiveness of the filter circuitry. The ripple frequency depends upon the frequency of the line and the type of rectifier circuitry employed. The frequency for a halfwave power supply, for example, equals the line frequency, while the ripple frequency for a full-wave type circuit equals twice the line frequency.

(F) *Voltage Regulations*

All power supply output potentials decrease as additional current is taken. A design and test factor equation used to determine the percent of voltage regulation is based upon maximum and minimum current usage. The formula is:

$$\%R = \frac{E_n - E_f}{E_f} \times 100$$

$\%R$ = regulation percentage
E_n = no load voltage (minimum current)
E_f = full load voltage (maximum current)

3.16-2   Half-Wave—One Diode—Transformer Type

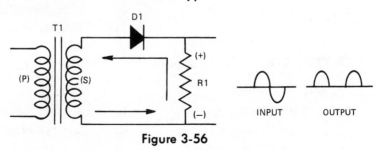

Figure 3-56

1. TRANSFORMER (T_1) SPECIFICS

(A) *Primary Current*

$$I_p = 1.57 \times \frac{I_{av}E_s}{E_p}$$

I_p = primary line current
I_{av} = average load current
E_s = secondary voltage
E_p = primary voltage

(B) *Primary or Secondary Volt-Ampere Ratings*

$$VA = 3.49 \times P$$

VA = volt ampere rating (rms voltage × rms current)
P = DC output wattage

(C) *Secondary Current*

$$I_s = 1.57 \times I_{AV}$$

I_s = secondary line current
I_{AV} = average output current

(D) *Secondary Voltage*

$$E_s = 2.22 \times E_{av}$$

E_s = secondary voltage
E_{av} = average output voltage

2. RECTIFIER (D_1) SPECIFICS

(A) *Peak Inverse Voltage Rating*

$$P_{IV} = 3.14 \times E_{av}$$
$$\text{or}$$
$$P_{IV} = 1.41 \times E_s$$

P_{IV} = Peak inverse voltage
E_{av} = average output voltage
E_s = secondary rms voltage

(B) *Peak Current Value*

$$I_D = 3.14 \times I_{av}$$

I_D = peak current of diode
I_{av} = average output current

(C) *rms Current Value*

$$I_D = 1.57 \times I_{AV}$$

I_D = rms current of diode
I_{AV} = Average output current

3. OUTPUT VOLTAGE (E_{R1}) SPECIFICS

(A) *rms Voltage*

$$E = 1.57 \times E_{AV}$$

E = rms output voltage
E_{AV} = average output voltage

(B) *Peak Voltage*

$$E_p = 3.14 \times E_{av}$$

E_p = peak output voltage
E_{av} = average output voltage

4. RIPPLE FREQUENCY

$$F = F_s$$

F = ripple frequency
F_s = source or line frequency

3.16-3   Full Wave—Two Diode—Transformer Type

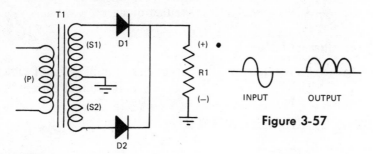

Figure 3-57

1. TRANSFORMER (T_1) SPECIFICS

(A) *Primary Current*

$$I_p = \frac{I_{av} \times E_{s1}}{E_p}$$

I_p = primary current
I_{av} = average output current
E_{s1} = onehalf secondary voltage
E_p = primary voltage

(B) *Primary Volt-Ampere Rating*

$$V_a = 1.11 \times P$$

V_a = volt-ampere rating
    (rms voltage times rms current)
P = output DC wattage

(C) *Secondary Volt-Ampere Rating*

$$V_a = 1.57 \times P$$

V_a = volt-ampere rating
P = dc output power

(D) *Secondary Current*

$$I_s = 0.707 \times I_{av}$$

I_s = secondary current
I_{av} = average output current

(E) *Secondary rms Voltage*

$$E_s = 2.22 \times E_{av}$$

E_s = total secondary voltage
E_{av} = average output voltage

2. RECTIFIER (D_1, D_2) SPECIFICS

(A) *Peak Inverse Voltage Rating*

$$P_{IV} = 3.14 \times E_{av}$$
or
$$P_{IV} = 2.82 \times E_{s1}$$

P_{IV} = peak inverse voltage
E_{av} = average output voltage
E_{s1} = secondary rms voltage of ½ winding

(B) *Peak Current Value*

$$I_D = 1.57 \times I_{av}$$

*I_D = peak current value of diode
I_{av} = average output current
*either diode

(C) *rms Current Value*

$$I_D = 0.785 \times I_{av}$$

I_D = rms current value of either diode
I_{av} = average output current

3. OUTPUT VOLTAGE (E_{R1}) SPECIFICS

(A) *rms Voltage*

$$E = 1.11 \times E_{av}$$

E = rms output voltage
E_{av} = average output voltage

(B) *Peak Voltage*

$$E_p = 1.57 \times E_{av}$$

E_p = peak output voltage
E_{av} = average output voltage

4. RIPPLE FREQUENCY

$$F = 2 F_s$$

F = ripple frequency
F_s = source or line frequency

3.16-4   Full Wave—4 Diode Bridge—Transformer Type

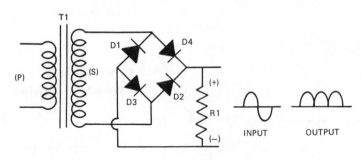

Figure 3-58

1. TRANSFORMER (T_1) SPECIFICS

(A) *Primary Current*

$$I_p = \frac{I_{av} \times E_s}{E_p}$$

I_p = primary current
I_{av} = average output current
E_s = secondary voltage
E_p = primary voltage

(B) *Primary and Secondary Volt-Ampere Rating*

$$V_a = 1.11 \times P$$

V_a = volt-ampere rating
     (rms voltage times rms current)
P = output DC wattage

(C) *Secondary Current*

$$I_s = I_{av}$$

I_s = secondary current
I_{av} = average output current

(D) *Secondary rms Voltage*

$$E_s = 1.11 \times E_{av}$$

E_s = secondary voltage
E_{av} = average output voltage

2. RECTIFIER (D_1, D_2, D_3, D_4) SPECIFICS

(A) *Peak Inverse Voltage Rating*

$$P_{IV} = 1.57 \times E_{av}$$
$$\text{or}$$
$$P_{IV} = 1.41 \times E_s$$

P_{IV} = peak inverse voltage
E_{av} = average output voltage
E_s = rms secondary voltage

(B) *Peak Current Value*

$$I_D = 1.57 \times I_{av}$$

I_D = peak current value of any diode
I_{av} = average output current

(C) *rms Current Value*

$$I_D = 0.785 \times I_{av}$$

I_D = rms current value of any diode
I_{av} = average output current

3. OUTPUT VOLTAGE (E_{R1}) SPECIFICS

(A) *rms Voltage*

$$E = 1.11 \times E_{av}$$

E = rms output voltage
E_{av} = average output voltage

(B) *Peak Voltage*

$$E_p = 1.57 \times E_{av}$$

E_p = peak output voltage
E_{av} = average output voltage

4. RIPPLE FREQUENCY

$$F = 2 \times F_s$$

F = ripple frequency
F_s = source or line frequency

3.16-5   Filters

Power supply filter circuits, designed to provide a practically unvarying DC potential by smoothing out pulsation, are classified into two groups. One group is called *capacitor input* type and the other a *choke input* type. Other terms relating to filter component configuration, such as *"pi" type, "L" type,* or *"T" type*, add variety and other filtering to the two main input capacitor and input choke filter classifications.

1. CAPACITOR INPUT TYPES

(Chokes may be used in place of resistors)

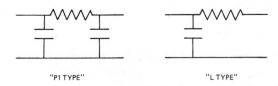

"P1 TYPE"   "L TYPE"

Figure 3-59

(A) *Characteristics:*
Output voltage is high but poor regulation

(B) *Input Capacitor Working Voltage*

$$E_c = 1.41 \times E_s$$

E_c = working voltage at light or no load
*E_s = second voltage

*Use all of secondary for halfwave or full-wave center tapped transformer types of power supplies. Use one-half of secondary for full-wave bridge power supply.

2. INDUCTOR INPUT TYPES

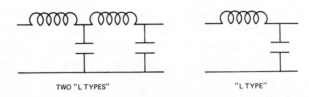

TWO "L TYPES"   "L TYPE"

Figure 3-60

(A) *Characteristics:*

Low output voltage—good regulation

(B) *Output Voltage*

$$E_0 = \frac{0.9\ E_T - (I_b + I_L)(R_1 + R_2) - E_R}{1000}$$

E_T = rms voltage applied to rectifiers
E_0 = output voltage
I_b = bleeder current
I_l = load current
R_1, R_2 = choke resistance (DC)
E_R = voltage dropped across rectifier

(C) *Ripple Value*

$$\%R = \frac{100}{LC}$$

$\%R$ = ripple for single filter section with 120 hertz frequency
L = value of inductor (henrys)
C = value of capacitor (farads)

Note: to reduce ripple to 5 percent or less, the product of LC must be 20.

$$\%R = \frac{650}{L_1 L_2\ (C_1 + C_2)^2}$$

$\%R$ = ripple for double filter section with 120 hertz
L_1, L_2 = value of inductors
C_1, C_2 = value of capacitors

(D) *Value of Choke*

$$L = \frac{RL}{1000}$$

L = value of choke for 120 hertz filter (henrys)
RL = load resistance (ohms)

$$L_1 = \frac{RL}{1000} \times \frac{120}{F}$$

L_1 = value of choke for other frequencies
RL = load resistance
F = actual frequency value

3.16-6   Power Supply Troubleshooting Symptoms

Locating circuitry malfunctions can be simplified by using component cause-effect analysis. The chart following outlines specific facts which relate to power supply components which might become defective. The chart is based upon two extremes, one of components being shorted and the other of being open. The letter abbreviation usage is as indicated here:

NOR = normal

LTN = lower than normal

GTN = greater than normal

PK (in) = peak of input DC voltage

PK (sec) = peak of secondary DC voltage

IN (F) = input frequency

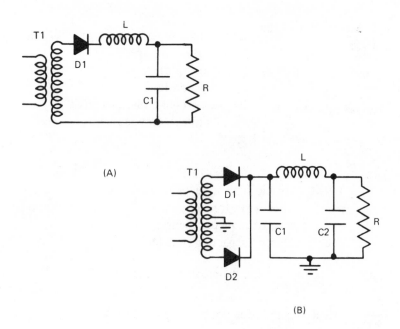

(A)

(B)

Figure 3-61

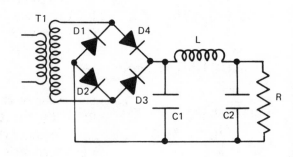

(C)

Defective Component	OUTPUT VOLTAGE			PEAK INVERSE VOLTAGE			RIPPLE AMPLITUDE			RIPPLE FREQUENCY		
	(A)	(B)	(C)	(A)	(B)	(C)	(A)	(B)	(C)	(A)	(B)	(C)
T_1 pr/sec open	OV.	OV.	OV.	OV.	OV.	OV.	0	0	0	0	0	0
T_1 pr/sec shorted	OV.	OV.	OV.	OV.	OV.	OV.	0	0	0	0	0	0
½ T_1 sec in Fig. B opens/shorts	—	½ NOR	—	—	NOR for 1 diode	—	—	GTN	—	—	½ NOR	—
D_1 or D_2 or D_3 or D_4 opens	OV.	LTN	filtered ½ wave	PK (in)	NOR	PK (sec)	None	GTN	GTN	None	½ NOR	Equals in (F)
D_1 or D_2 or D_3 or D_4 shorts	ACV.	Low AC	filtered ½ wave	None	NOR for good diode	PK (sec)	Equals AC	Pk to Pk of output AC	GTN	Equals AC	Equals IN (F)	Equals IN (F)
D_1 and D_2 or D_3 and D_4 opens/shorts	—	—	OV.	—	—	None	—	—	None	—	—	None
"L" opens	OV.	OV.	OV.	None	NOR	None	GTN	None	None	None	None	None
"L" shorts	GTN	GTN	GTN	2 × PK input	NOR	PK (sec)	GTN	GTN	GTN	IN (F)	NOR	2 × PK input
"C_1" opens	LTN	LTN	LTN	PK (in)	NOR	PK (sec)	GTN	GTN	GTN	IN (F)	NOR	2 × PK input
"C_1" shorts	OV.	OV.	OV.	PK (in)	½ NOR	PK (sec)	None	None	None	None	None	None
"C_2" opens	—	LTN	LTN	—	NOR	NOR	—	GTN	GTN	—	NOR	NOR
"C_2" shorts	—	OV.	OV.	—	½ NOR	½ NOR	—	None	None	—	None	None
R opens	PK of DC	PK of DC	PK of DC	2 × DC PK	NOR	PK (sec)	None	None	None	None	None	None
R shorts	OV.	OV.	OV.	PK (in)	½ NOR	PK (sec)	None	None	None	None	None	None

3.17   REACTANCE (AC APPLICATION ONLY)

The opposition to alternating current offered by circuitry having inductance or capacitance is called reactance. The letter symbol used to identify like grouping of AC oppositions is the letter "X." Whenever unlike AC oppositions are encountered in the same circuitry, the total opposition is termed impedance and is represented by the letter "Z." The important formulas and data regarding reactance are outlined in this section of chapter 3.

3.17-1   Formulas

(A) *Capacitive Reactance*

$$X_C = \frac{1}{2\pi fC}$$

$$X_C = \frac{1}{\omega C}$$

$$X_C = \frac{0.159}{fC}$$

X_C = capacitive reactance (ohms)
f = applied frequency (hertz)
C = unit of capacitance (farads)
$\omega = 2\pi f$

(B) *Inductive Reactance*

$$X_L = 2\pi fL$$

$$X_L = \omega L$$

X_L = inductive reactance (ohms)
f = frequency (hertz)
L = value of inductor (henrys)

3.17-2   Series

(A) *Same kind*

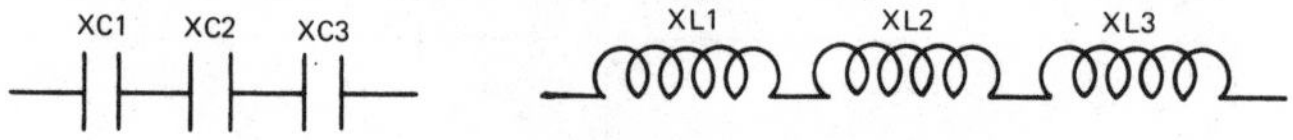

Figure 3-62

General $X_T = X_1 + X_2 + X_3 \ldots .$ X_T = total reactance (ohms)
Capacitors $X_{CT} = X_{C1} + X_{C2} + X_{C3} \ldots .$ $X_1 - X_3$ = individual reactances (ohms)
Inductors $X_{LT} = X_{L1} + X_{L2} + X_{L3} \ldots .$

Note: The letter "Z" for impedance may be substituted for either X_C or X_L since a combination of reactances is possible.

(B) *Opposite Kind*

Figure 3-63

$$Z \text{ (or } X_T) = X_L - X_C$$

$$Z \text{ (or } X_T) = X_C - X_L$$

X_T = total reactance (ohms)

Z = total impedance (ohms)

X_L = inductive reactance (ohms)

X_C = capacitive reactance (ohms)

Note: resultant reactance is smaller than largest individual reactance.

3.17-3   Parallel

(A) *Same kind—two only*

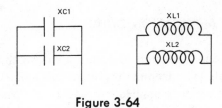

Figure 3-64

Note

$$X_{CT} = \frac{1}{2 \pi f (C_1 + C_2)}$$

$$X_{LT} = 2 \pi f \left(\frac{L_1 L_2}{L_1 + L_2} \right)$$

$$X_T = \frac{X_1 X_2}{X_1 + X_2}$$

X_T = total reactance (ohms)

X_1, X_2 = inductive or capacitive reactance of individual components (ohms)

(B) *Same kind—more than two*

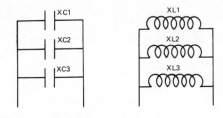

Figure 3-65

$$X_T = \frac{1}{\dfrac{1}{X_1} + \dfrac{1}{X_2} + \dfrac{1}{X_3} + \cdots}$$

X_T = total reactance (ohms)

X_1, X_3 = individual capacitive or inductive reactance of components (ohms)

(C) *Opposite Kind*

$$X_T = \frac{X_L X_C}{X_L - X_C}$$

X_T = total reactance (ohms)
X_C = capacitive reactance (ohms)
X_L = inductive reactance (ohms)

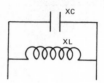

Figure 3-66

Note: resultant reactance is larger
than smallest reactance

3.18   RESONANCE (AC APPLICATION ONLY)

Tuned circuits have only two components, inductors and capacitors, which are connected in series, parallel, or series-parallel. The inherent opposition characteristics of the reactive components change when the frequency applied increases or decreases. A condition called resonance exists when a specific frequency appears and causes the tuned circuit to lose its reactive property. The resonant frequency is determined by the value of capacitance and inductance in the circuit.

3.18-1   Resonance Formula

$$f_r = \frac{1}{2 \pi \sqrt{LC}}$$

$$f_r = \frac{0.159}{\sqrt{LC}}$$

f_r = resonant frequency (hertz)
L = value of inductor (henrys)
C = value of capacitor (farads)

3.18-2   Quality Factor (Q or Figure of Merit)

(A) *Series resonant circuit*

$$Q = \frac{X}{R_S}$$

Q = quality factor
R_S = series resistance (ohms)
X = reactance (X_L or X_C in ohms)

(B) *Parallel Resonant Circuit*

$$Q = \frac{Z_R}{X}$$

Q = quality factor
Z_R = resistive impedance at resonance (ohms)
X = reactance (X_L or X_C in ohms)

3.18-3   Series and Parallel "LCR" Summary

	SERIES "LCR" CIRCUIT			PARALLEL "LCR" CIRCUIT		
	Above fr	fr	Below fr	Above fr	fr	Below fr
Circuit appearance	XL+R	R	XC+R	XC+R	R	XL+R
Circuit current	low	high	low	high	low	high
Circuit impedance	high	low	high	low	high	low
Current phase	lags voltage	in phase	leads voltage	leads voltage	in phase	lags voltage

fr = resonant frequency
XL = inductive reactance
XC = capacitive reactance

3.19   TIME CONSTANTS (AC APPLICATION ONLY)

Circuitry having resistors, capacitors, and inductors exhibit charge and discharge time constant characteristics. Normally, these characteristics are based upon the circuit's ability to change from one steady state condition to another, simply termed DC-transient periods. The two general circuitry groups are resistor-capacitor (R-C) and resistor-inductor (R-L).

3.19-1   R-C Circuits

(A) *General Formula*

$$TC = RC$$

TC = time constant (seconds)
R = value of resistance (ohms)
C = value of capacitance (farads)

Resistance Unit	Capacitance Unit	Time Constant Unit
ohms	Farads	Seconds
Megohms	Microfarads	Seconds
Ohms	Microfarads	Microseconds
Megohms	Picofarads	Microseconds

The time constants required for the capacitor illustrated in Figure 3-67 to become fully charged (5-time constants) or discharged as the case may be, are as shown: the calculations were based upon Figure 3-67 circuitry values, and time "0" represents the instant the switch is thrown.

Formula

$Tc = RC$
$Tc = 1 + 10^{+5} (0.01 \times 10^{-6})$
$*Tc = 0.001 \text{ sec} = 1 \text{ millisecond}$
*one time constant (requires five
 time constants for full charge).

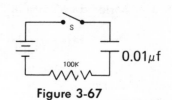

Figure 3-67

Time Constant	Time (milliseconds)	Ec (volts)	ER (volts)	Current (MA)
0	0	0	100.00	10.00
1	1	63.20	36.80	3.680
2	2	86.46	13.54	1.354
3	3	95.02	4.98	0.498
4	4	98.17	1.83	0.183
5	5	99.32	0.68	0.068
6	6	99.75	0.25	0.025

3.19-2   R-L Circuits

(A) *General formula*

$$Tc = \frac{L}{R}$$

Tc = time constant (seconds)
L = value of inductor (henrys)
R = value of resistor (ohms)

3.20   TRANSFORMER FORMULAS (AC APPLICATION ONLY)

Transformers transfer alternating voltage from the primary to secondary and can provide the same voltage on its secondary as on its primary, or the secondary voltage can be stepped-up or down depending upon circuitry requirements. The symbols used in the formulas describing transformer facts are listed first to avoid duplication.

Voltage

Ep = primary
Es = secondary

Currents

Ip = primary
Is = secondary

Power

Po = output power
Pi = input applied power

Impedances

Zp = primary
Zs = secondary
Za = component in series with primary
Zb = component in series with secondary
Zr = ratio

Efficiency

Eff = efficiency factor (decimal
N = efficiency factor

Turns

T = primary to secondary turns ratio
N_p = turns of primary
N_s = turns of secondary

3.20-1   Voltage

(A) *Primary*

$$E_p = \frac{E_s N_p}{N_s} \qquad \text{OR} \qquad E_p = \frac{E_s I_s}{I_p}$$

(B) *Secondary*

$$E_s = \frac{E_p N_s}{N_p} \qquad \text{OR} \qquad E_s = \frac{E_p I_p}{I_s}$$

3.20-2   Current

(A) *Primary*

$$I_p = \frac{E_s I_s}{E_p} \qquad \text{OR} \qquad I_p = \frac{N_s I_s}{N_p}$$

(B) *Secondary*

$$I_s = \frac{E_p I_p}{E_s} \qquad \text{OR} \qquad I_s = \frac{N_p I_p}{N_s}$$

3.20-3   Turns

(A) *Primary*

$$N_p = \frac{E_p N_s}{E_s} \qquad \text{OR} \qquad N_p = \frac{N_s I_s}{I_p}$$

(B) *Secondary*

$$N_s = \frac{E_s N_p}{E_p} \qquad \text{OR} \qquad N_s = \frac{N_p I_p}{I_s}$$

(C) *Step-up Turns Ratio*

$$T = \frac{N_s}{N_p}$$

(D) *Step-down Turns Ratio*

$$T = \frac{N_p}{N_s}$$

3.20-4 Impedance

(A) *Ratio*

$$\frac{Z_p}{Z_s} = \left(\frac{N_p}{N_s}^2\right) \qquad OR \qquad Z_p = Z_s T^2$$

(B) *Step-up Ratio*

$$Z_s = Z_r Z_p$$

(C) *Step-down Ratio*

$$Z_s = \frac{Z_p}{Z_r}$$

(D) *Total—looking into primary*

$$Z_t = Z_a + Z_p - \frac{Z_m^2}{Z_b + Z_s}$$

3.20-5 Efficiency

(A) *General Formula*

$$Eff = \frac{P_o}{P_I}$$

(B) *Power*

$$P_o = N\ P_I \qquad OR \qquad P_p = Eff\ P_I$$

3.20-6 Voltage, Current, and Turn Relationships

$$\frac{E_p}{E_s} = \frac{N_p}{N_s} = \frac{I_s}{I_p}$$

3.21 TRANSISTOR SPECIFICS

Although there are hundreds of different transistor types employed throughout the electronics industry, only three main circuit configurations are commonly used. These three configurations include the grounded or common base, the grounded or emitter follower or common collector, and the grounded or common emitter. In all three of these circuit configurations, the base element is always one of the two input terminals and the collector is always one of the two output terminals.

3.21-1 Common Base

Current gain for the common base configuration uses an expression termed alpha (α), however there is an AC gain and a DC gain. The AC alpha is represented as h_{fb} while DC alpha is h_{FB}. The formulas for each are:

(A) *AC alpha* = ratio between a change in collector current and a change in emitter current *with a fixed collector voltage*.

$$\alpha = h_{fb} = \frac{\Delta\ Ic}{\Delta\ Ie}$$

$\Delta\ Ic$ = change in collector rms current

$\Delta\ Ie$ = change in emitter rms current

(B) *DC alpha* = ratio between collector DC current and the emitter DC current.

$$\alpha = h_{FB} = \frac{IC}{IE}$$

IC = collector DC current

IE = emitter DC current

Note: Alpha for the common base configuration is always less than 1.0 but can approach unity. This type of circuitry provides phase inversion of the output signal with respect to the input.

3.21-2   Common Collector

The common collector configuration never has a voltage gain greater than one. The formulas describing common collector specifics are:

(A) *Voltage Gain*

$$Av = 1$$

Av = voltage gain

(B) *Current and Power Gain*

$$G = Ai = \frac{1}{1 - \alpha}$$

G = power gain

Ai = current gain

α = alpha

Note: The common collector exhibits the highest input resistance in comparison to the common base or common emitter configurations.

3.21-3   Common Emitter

The common emitter configuration is more frequently used because it offers the greatest gain for all values of load resistance. The gain, described by BETA (β), is a ratio between the base and collector currents. AC BETA is represented by h_{fe} while DC BETA is h_{FE}. Beta values are greater than one and often run into the hundreds.

(A) *AC BETA* = ratio between a change in collector current and a change in base current *while a fixed collector voltage is applied*.

$$\beta = h_{fe} = \frac{\Delta Ic}{\Delta\ Ib}$$

$\Delta\ Ic$ = change in collector rms current

$\Delta\ Ib$ = change in base rms current

(B) *DC BETA* = ratio between DC base current and the DC collector current.

$$\beta = h_{FE} = \frac{IC}{IB}$$

IC = DC collector current

IB = DC base current

3.21-4   Alpha-Beta Relationship

$$\alpha = \frac{\beta}{\beta + 1}$$

α = common base gain

β = common emitter gain

$$\beta = \frac{\alpha}{1 - \alpha}$$

3.21-5   Input Resistance

$$R_i = \frac{\Delta V_i}{\Delta I_i}$$

R_i = input resistance
V_i = source or input voltage
I_i = input current

3.21-6   Output Resistance

$$R_o = \frac{\Delta V_o}{\Delta I_o}$$

R_o = output resistance
V_o = output voltage
I_o = output current

3.21-7   Power Gain

$$A_p = \frac{\Delta P_o}{\Delta P_i}$$

A_p = power gain
P_o = output power
P_i = input power

3.21-8   Voltage Gain

$$A = \frac{\Delta V_c}{\Delta V_b}$$

A = voltage gain
V_c = collector voltage
V_b = base voltage

*collector current held constant

3.22   VACUUM TUBE AMPLIFIER SPECIFICS

Amplifiers can be classified in three general ways. The first classification involves frequency of operation plus the necessary coupling circuitry. Amplifiers found within this group include DC, audio, video, and RF amplifiers. The second classification separates vacuum tube circuitry into three groups: voltage, current and power amplifiers. The third classification describes the output reproduced signal. Examples of this classification include class A, AB, B, and C amplifiers.

3.22-1   Characteristic Curves

Each vacuum tube has its own set of curves which denote important factors during circuitry design steps. Two such curves are illustrated in Figures 3-68A and 3-68B. Curve ''A'' (eg-Ip) illustrates grid voltages to plate current relationships. This particular curve results when no plate load resistor is employed and when the plate voltage is held constant during the individual plate voltage curve values. The Ep-Ip curve shown in 3-68B results when the grid voltage is held constant during individual grid voltage curve values.

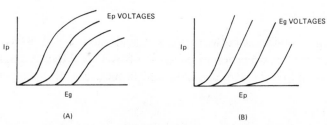

Figure 3-68

3.22-2   Load Lines

Before selecting the amount of bias necessary to make the vacuum tube amplifier circuitry a particular class of operation, a DC load line must be drawn. The three things required to do this are: a set of Ep-Ip curves for the tube, knowledge of plate voltage value, and the amount of plate load resistance to be used.

To demonstrate this concept we will say B+ is 300 volts, RL is 50,000 ohms, and the family of Ep-Ip curve for the tube resembles the one shown in Figure 3-69.

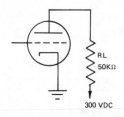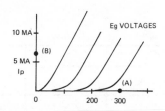

Figure 3-69

First: Determine I_p (point B)

$$I_p = \frac{E_s}{RL}$$

$$I_p = \frac{300V}{50k\ \Omega}$$

$$I_p = \ 6\ mA$$

I_p = plate current (maximum value)
E_s = source or applied voltage
RL = value of load resistor

*Assume tube has no resistance

Second: Assume no voltage drop across plate load resistor. Point "A" represents this potential of 300 volts. The DC load line is constructed by drawing a straight line between point "B" (maximum current) and point "A" (maximum voltage). The grid voltage lines crossing the load line serve as a reference for tube biasing.

3.22-3   Classes of Bias

(A) *Class A, A_1, or A_2*

This classification means the vacuum tube is biased in the linear portion of the family curve, thereby providing the highest possible fidelity. Subscript 1 indicates no grid current flowing, subscript 2 means the biasing is the same as for A_1 but grid current flows.

(B) *Class AB*

Biasing is midpoint between class A and class B just above the cutoff point which is not on the linear portion of the curve. If AB_1 or AB_2 were indicated, then grid current flow would be the same as indicated in the class A explanation.

(C) *Class B*

Biasing is at the cutoff point. Grid current flows.

(D) *Class C*

Biasing is well below the cutoff point. Grid current flow is normally indicated by the C_2 designation.

(E) *Amplifier Classification Summary*

Class	Input	Output	Conduction Percentage	Efficiency	Fidelity
A	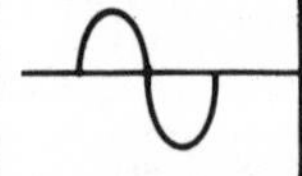	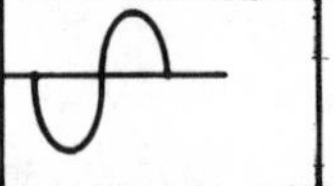	100%	Low	High
AB	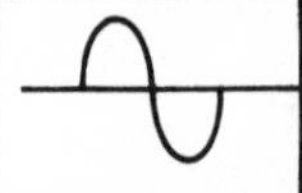	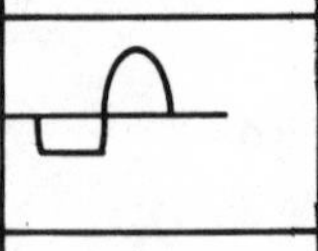	Less than 100% More than 50%	Medium	Medium
B	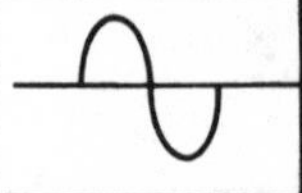	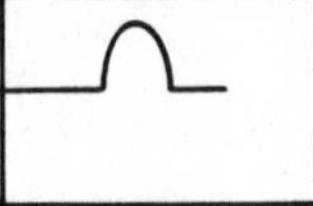	50%	High	Low
C	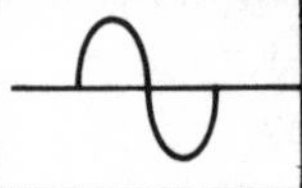	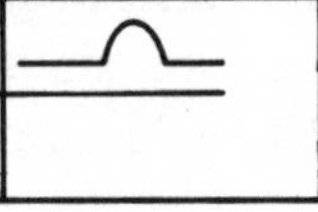	Less than 50%	Highest	Lowest

Figure 3-70

3.22-4   Amplification Factor

(A) $\mu = (Gm)(rp)$

(B) $\mu = \dfrac{\Delta\ Eb}{\Delta\ Ec}$ with Ib constant

(C) $\mu = \dfrac{\Delta\ Ep}{\Delta\ Eg}$ with Ip constant

μ = amplification factor
Eb, Ep = plate voltage
Eg, Ec = grid voltage
Ip, Ib = plate current
Gm = transconductance
rp = plate resistance (dynamic)

3.22-5   Transconductance

(A) $Gm = \dfrac{\mu}{rp}$

(B) $Gm = \dfrac{\Delta\, Ip}{\Delta\, Eg}$

*Gm = transconductance (mutual conductance)
μ = amplification factor
rp = AC plate resistance
Ip = plate current
Eg = grid voltage

*Tells ease with which current flows-measured in siemans (s), formerly mhos.

3.22-6   Plate Resistance

(A) *DC Resistance*

$Rp = \dfrac{Eb}{Ib}$

Rp = DC plate resistance
Eb = plate to cathode voltage (no signal)
Ib = plate current (no signal)

(B) *AC Resistance*

$rp = \dfrac{\Delta\, Ep}{\Delta\, Ip}$ with eg constant

*rp = AC (dynamic) plate resistance
Ep = alternating plate voltage
Ip = alternating plate current
eg = instantaneous grid voltage

*Computed from several readings—grid voltage held constant

3.22-7   Cathode Resistor

$Rk = \dfrac{Zorp}{rp - Zo\,(\mu + 1)}$

OR

$Rk = \dfrac{Eg}{Ik}$

Rk = cathode resistor
Zo = output impedance
rp = AC plate resistance
μ = amplification factor
Eg = grid voltage
Ik = cathode current

3.22-8   Output Impedance

(A) *Cathode Follower Tube*

$Zo = \dfrac{Rk\, rp}{rp + Rk\,(\mu + 1)}$

Zo = output impedance
Rk = cathode resistance
rp = AC plate resistance
μ = amplification factor

(B) *High-mu tube*

$$Zo = \frac{Rk}{1 + RkGm}$$

Zo = output impedance
Rk = cathode resistor
Gm = transconductance

3.22-9   Gain

(A) *General*

$$A = \frac{Eout}{Ein} = \frac{Eo}{eg}$$

OR

$$A = \frac{\mu\ Rk}{rp + Rk\ (\mu + 1)}$$

OR

$$A = \frac{\mu\ ZL}{rp + ZL}$$

A = vacuum tube gain
Eo = output voltage
eg = input signal voltage
μ = amplification factor
rp = AC plate resistance
Rk = cathode resistor
ZL = equivalent load impedance/plate load

(B) *Current in dB*

$$dB = 20 \log \frac{Io}{Ii}$$

dB = decibel
Io = output current
Ii = input current

(C) *Voltage in dB*

$$dB = 20 \log \frac{Eo}{Ei}$$

dB = decibel
Eo = output voltage
Ei = input voltage

(D) *Power in dB*

$$dB = 10 \log \frac{Po}{Pi}$$

dB = decibel
Po = output power
Pi = input power

(E) *Stages in cascade*

$$AT = A_1 \times A_2 \times A_3$$

OR

$$dBT = dB_1 + dB_2 + dB_3$$

AT = total voltage gain
A₁-A₃ = individual stage voltage gains
dBT = total decibel gain
dB₁-dB₃ = individual stage power gains

(F) *Feedback amplifiers*

$$AT = \frac{Eo}{Ei}$$

OR

$$AT = \frac{A}{1-BA}$$

AT = total gain
Eo = output voltage
Ei = input voltage
A = gain without feedback
B = fraction of output feedback

(G) *High-mu tubes*

$$A = \frac{R_kG_m}{1 + R_kG_m}$$

OR

$$A = Z_oG_m$$

A = gain
R_k = cathode resistance
G_m = transconductance
Z_o = output impedance

3.22-10   Video Amplifier Frequency Bands

(A) *High Band Passed*

$$FH = \frac{1}{2\,\pi\,R_LC_S}$$

FH = highest frequency in band passed
RL = total load resistance
Cs = total shunt capacitance

(B) *Lowest Band Passed*

$$FL = \frac{1}{2\,\pi\,R_gC_c}$$

FL = lowest frequency in band passed
Rg = grid resistor value
Cc = coupling capacitor value

3.23   WAVELENGTH AND WAVEGUIDE SPECIFICS

Different techniques, formulas, and application are necessary when high frequency, namely frequencies in the microwave region, is employed. At this high frequency, component and circuitry characteristics differ greatly from those encountered in low frequency application. This section of Chapter 3 outlines those facts which adapt themselves readily to high frequency application.

3.23-1   Velocity of Propagation

(A) *Meters*

$$\lambda = \frac{300,000}{F}$$

OR

$$\lambda = \frac{300}{f}$$

λ = wavelength in meters
F = frequency in kilohertz
f = frequency in megahertz

(B) *Feet*

$$\lambda = \frac{300{,}000(3.28)}{F}$$

OR

$$\lambda = \frac{984{,}000}{F}$$

OR

$$\lambda = \frac{984}{f}$$

λ = wavelength in feet
F = frequency in kilohertz
f = frequency in megahertz

3.23-2   Velocity Factor

$$V_F = \frac{V_m}{V_s}$$

*V_F = velocity factor
V_m = velocity of material (air, etc.)
V_s = velocity in space

*Always less than one

3.23-3   Standing Wave Ratio

$$*SWR = \frac{Z_R}{Z_o}$$

OR

$$*SWR = \frac{Z_o}{Z_R}$$

OR

$$SWR = \frac{I_{max}}{I_{min}}$$

OR

$$SWR = \frac{E_{max}}{E_{min}}$$

SWR = standing wave ratio
Z_R = impedance of load (pure resistance)
Z_o = characteristic impedance of line
I = maximum or minimum value of current
E = maximum or minimum value of voltage

*Larger value is the numerator

3.23-4   Waveguides with TE_{01} Modes

In either the transverse electric (TE) or transverse magnetic (TM) modes of operation, subscripts are used to define the mode. An example would be TE_{mn} where the "m" represents the first subscript and the letter "n" the second. Figure 3-71 should be referred to during waveguide-formula applications.

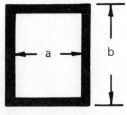

Figure 3-71

(A) *Characteristic Impedances*

$$Z_0 = \frac{465(a)}{b\sqrt{1 - \frac{(\lambda)^2}{(2b)}}}$$

OR

$$Z_0 = \frac{930(a)}{\sqrt{4b^2 - (\lambda)^2}}$$

OR

$$*Z_0 = 120\pi \frac{\lambda g}{\lambda}$$

Z_0 = characteristic impedance
a = short waveguide dimension
b = long waveguide dimension
λ = wavelength in space (air)
λg = wavelength in waveguide

*For TE and Tm modes

(B) *Waveguide dimensions*

$$b = 0.7\ (\lambda)$$
$$a = 0.35\ (\lambda)$$

b = long waveguide dimension
a = short waveguide dimension
λ = wavelength in space

(C) *Sine and Cosine Functions for Wave Fronts*

1. $\text{Sine } \Theta = \dfrac{\lambda}{\lambda g}$

2. $\text{Sine } \Theta = \dfrac{V g}{V a}$

3. $\text{Sine } \Theta = \dfrac{V a}{V p}$

4. $\text{Cos } \Theta = \dfrac{\lambda}{2b}$

Sin = trignometric sine function
Cos = trignometric cosine function
Θ = angle wave front makes with waveguide wall
λ = wavelength in space (air)
λg = wavelength in waveguide
$V g$ = velocity in waveguide
$V p$ = phase velocity
b = long waveguide dimension
a = short waveguide dimension

(D) *Wavelength*

1. *(at cutoff)*

$$\lambda c = \frac{2}{\sqrt{\left(\dfrac{m}{a}\right)^2 + \left(\dfrac{n}{b}\right)^2}}$$

OR

$$\lambda c = 2b$$

λ = wavelength in space

λg = wavelength in waveguide

λc = cutoff wavelength

m = first subscript in "TE" or "TM" mode designations

n = second subscript in "TE" or "TM" mode designations

b = long waveguide dimension

a = short waveguide dimension

2. *In Free Space*

$$\lambda = \frac{\lambda g}{\sqrt{1 + \left(\dfrac{\lambda g}{2b}\right)^2}}$$

OR

$$\lambda = \frac{(2b)(\lambda g)}{\sqrt{4b^2 + \lambda g^2}}$$

3. *In Waveguide*

$$\lambda g = \frac{\lambda}{\sqrt{1 - \left(\dfrac{\lambda}{2b}\right)^2}}$$

OR

$$\lambda g = \frac{2b\,\lambda}{\sqrt{(4b)^2 - (\lambda)^2}}$$

CHAPTER 4

Troubleshooting Electronic Components

4.1   CAPACITORS

Capacitors are found in all sizes and shapes, all ranges of capacitance and temperature coefficients, and all types of dielectric materials. All of these facts, plus more, contribute to capacitor reputations; reputations established during standardized, agency-approved testing and in actual circuitry operation. The reputation established for each different capacitor type provides for three easy general classification groupings to exist. One classification is low loss—good stability, another is medium loss—medium stability, and the third is electrolytics. Capacitors found within the low loss —good stability and medium loss—medium stability groups are non-polarized. It will not make any great electrical difference as to which end is connected where unless circuitry shielding is important. If shielding is desired, keep the band marked end at the negative most part of its circuit function.

4.1-1   Low Loss—Good Stability

Capacitors grouped within this classification include mica, glass, some ceramics, and plastics. The variations within each specific type allow choices of manufactured capacitors because they are: small, temperature stable, tubular, disc, multiple sectioned, or a substrate, a plug in, a monolithic, a trimmer, a feed-through or has spark gaps, axial leads, specific or general purpose applications.

It would be almost impossible to cover all characteristics pertinent to every low loss-good stability capacitor type, but the following chart-type data provides the capacitive value and normal dc operating voltages for each type listed.

4-1A   LOW LOSS—GOOD STABILITY (MICA)

Capacitance Range—1pF-100,000pF
Voltage Range—50V to 35K VDCW

Normally, if packaging is dipped or encapsulated, manufacturer identification, capacitor value in pF, voltage rating and temperature coefficient are stamped on the capacitor if space allows. Sometimes letters and numbers are used; if so, the letters denote fixed mica-dielectric while the numbers describe dimensions and physical size. The standard tolerance is "J" (± 5%) while the minimum standard is "F" (± 1%).

When molded case styles of packaging are used, color code interpretation is necessary. Figure 4-1 illustrates which dot-color positioning means what.

Color	1st & 2nd Digit	Multiplier	Tolerance	Voltage	Characteristics	
Black	0	1	± 20%	—	A	NP0
Brown	1	10	± 1%	100V	B	N033
Red	2	100	± 2%	—	C	N075
Orange	3	1000	—	300V	D	N150
Yellow	4	10,000	—	—	E	N220
Green	5	—	± 5%	500V	F	N330
Blue	6	—	—	—		N470
Violet	7	—	—	—		N750
Gray	8	—	—	—	—	—
White	9	—	—	—	—	—
Gold	—	0.1	± ½%	—	—	—
Silver	—	0.01	± 10%	—	—	—

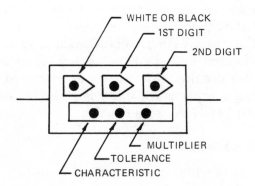

Figure 4-1

Characteristic	Temp. Coefficient	Voltage Rating	Tolerance	Operating Temp.
A (NP0)	—	100	—	—
B (N033)	—	250	—	—
C (N075)	−200 to +200	300	1pF	—
D (N150)	−100 to +100	500	0.5pF	—
E (N220)	−20 to +100	600	—	—
F (N330)	0 to +70	1,000	1%	—
G	—	1,200	2%	—
H	—	1,500	—	—
J	—	2,000	5%	—
K	—	2,500	10%	—
L	—	3,000	—	—
M	—	4,000	—	−55 to +70
N	—	5,000	—	−55 to +85

Characteristic	Temp. Coefficient	Voltage Rating	Tolerance	Operating Temp.
P	—	6,000	—	−55 to +150
Q	—	8,000	—	—
R	—	10,000	—	—
S	—	12,000	—	—
T	—	15,000	—	—
U	—	20,000	—	—
V	—	25,000	—	—
W	—	30,000	—	—
X	—	35,000	—	—

4.1-1B   GLASS/QUARTZ

Capacitive Range—50pF to 0.005mF
Voltage Range—1V to 300V

This particular type, mostly used for trimmers, may be rectangular in size or cylindrical in shape, having axial leads if rectangular, and special lead design if for panel mounting or printed circuit mounting. Typical temperature coefficients range from 0 to 140.

4.1-1C   CERAMICS

	Capacitive Range	Voltage Range
General Applications	5 pF to 2.5 mF	3V to 200V
High Voltage	100pF to 1000pF	10kV to 40kV
RF (transmitting)	3pF to 1000pF	5kV to 40kV
Feed through	0.5pF to 3000pF	100V to 600V

Markings for ceramic dielectric capacitors normally include company trademarks, capacitance value, temperature coefficient, voltage and tolerances on disc and tubular capacitors if space permits. Feed-through capacitors, being of the symmetrical, eyelet, bushing, or tubular variety, are normally marked by colors instead of printed data. A single band at one end of ceramic tubular capacitors indicates the inner electrode connection.

4.1-1D   PLASTICS

	Capacitive Range	Voltage Range
Polystyrene	0.01mF to 10mF	100V to 2kV
Polystrene-Polyester	0.001mF to 1mF	50V to 1kV

Generally speaking, capacitor type plastics within these capacitive ranges have excellent characteristics which are independent of applied frequencies. Packaging styles vary somewhat among manufacturers, but company trademarks, capacitances, tolerance and voltage ratings are usually printed on the capacitor body. The color-coded end, or black band end, normally identifies the outside foil lead, but this same marking has also been used for inner lead identification.

4.1-2   Medium Loss—Medium Stability

This grouping includes paper, plastic film, and high "K" ceramic capacitor types. The paper type has been largely replaced by plastic film varieties, but still is excellent for high AC and DC voltage applications. The plastic film types include polystyrene, polyester and polycarbonates, the latter being smaller in physical size. High "K" type ceramics, sometimes called temperature-compensating ceramics, were designed mainly for bypass and coupling applications since their use for these purposes would not affect normal circuitry operation.

4.1-2A   PAPER TYPES

	Capacitive Range	Voltage Range
General Application	0.001mF to 200mF	50V to 200kV
Paper-polyester	0.001mF to 30mF	100V to 15kV
Paper-metallized	0.001mF to 125mF	50V to 600V
Metallized paper-polyester	0.001mF to 20mF	200V to 600V

Paper types usually cost less than other capacitors suitable for the same applications. The physical size, however, may limit their use in electronic circuitry. Normally their shapes are tubular. Manufacturer identification, capacitance, voltage rating, and other possible markings will be found printed on their body.

4.1-2B   PLASTIC FILM TYPES

	Capacitive Range	Voltage Range
Polyester	0.001mF to 20mF	50V to 1kV
Metallized Polyester	0.01mF to 20mF	50V to 600V
Polycarbonate	0.001mF to 20mF	50V to 600V
Metallized Polycarbonate	0.01mF to 5mF	50V to 600V

Capacitor types using plastic film dielectrics have dipped, film wrap, molded, or ceramic encasements. Printed data on each capacitor identifies the manufacturer, capacitance and voltage range plus tolerances. Outer foil lead is usually identified by a color coded end or black line.

4.1-2C   HIGH "K" CERAMICS

	Capacitive Range	Voltage Range
General Application	1.5pF to 0.001mF	500V to 1000V

High "K" or temperature compensating capacitors usually are manufactured as disc types, having specific temperature change ratings. Temperature coefficients identification of "NP0" (negative, positive, zero) designate amount of temperature changes in parts-per-million-per-degree centigrade or ppm/°c. The letter "N" or a minus sign indicates a decrease in capacitance, while a "P" or plus sign means an increase. A positive temperature coefficient of 450 means the capacitance will increase

450/1,000,000 or 0.045 percent for each degree Celsius temperature rise. A negative temperature coefficient means a decrease in capacitance with a rise in temperature. Capacitance value, voltage, capacitive and temperature tolerances are usually found imprinted on the capacitor.

4.1-3   Electrolytics

Capacitors normally having the highest capacitance per volume unit rating are called electrolytics. This term is given to any capacitor whose dielectric layer is formed using electrolytic methods. Some of the capacitors within this group may not contain an electrolytic as such, but manufacturing processes common to both are used.

Two distinct types of electrolytic capacitors are commonly found in electronic circuitry, they are:

1. Aluminum oxide dielectic capacitors, being 99.9 percent reliable, have a shelf life from about 6 months to 5 years, after which reforming is necessary.

2. Tantalum oxide dielectric capacitors, smaller physically and not having shelf life limitations. All electrolytic capacitors are polarized and must be installed properly to avoid damage.

4.1-3A   ALUMINUM-OXIDE

Capacitive Range	Voltage Range
0.47mF to 1,000,000mF	2.5V to 7.50V

The electrolytic capacitors of the aluminum oxide dielectric type are obtainable with axial leads that are tubular or cylindrical in shape, having plug-in sockets, pins which can be soldered, (twistlock) or screw tightened connections. Cylindrical types can be single capacitors, double, triple, quadruple, or quintuple. At any rate, polarity of potentials must be observed; negative reference voltages to the negative leads and positive potentials to the positive leads.

Stamped or printed data found on electrolytic capacitors denote manufacturers by name or trademark, identification numbers used in stocking or catalog use, possible temperature rating, capacitance and voltage values plus polarity markings.

4.1-3B   TANTALUM-OXIDE

	Capacitance Range	Voltage Range
	0.15mF to 3500mF	2V to 300V
Tantalum (wet foil)	0.17mF to 500mF	4V to 150V
Tantalum (dry anode)	0.47mF to 330mF	6V to 35V

Tantalum dielectric electrolytic capacitors vary in size, shapes, and encasement styles. Some of the terms often encountered include tubulars, molded plastic, epoxy dipped, metal encased, hermetically sealed or foil types. Regardless of size or shape, markings printed on their body identify manufacturer, capacitance, voltage rating, tolerances and polarity. Often the negative most lead can be identified as being the lead attached to the metal sealed end, or as the lead being copper, or by a black band nearest the negative lead. In some marking systems, the positive lead is marked with plus (+) signs rather than pointing out the negative lead.

4.1-4   Capacitor Value Interpretation

Most capacitors present few problems when deciphering their capacitive value and working voltage rating. Ceramic types, paper types, plastic film types and electrolytic types all have capacitive values in microfarads. All mica types and some disc packaged type capacitors have capacitance values in micro-micro farads (mmF) commonly called pico farads (pF). The disc type capacitors generally can be separated into "mF" or "pF" after examining the printed data found on the capacitor. Normally, if a disc capacitor has a number equaling "1" or greater, the capacitor is rated in picofarads. If the disc has a number in decimal form less than one, the capacitor value is in micro-farads. If you were to examine micro-farad-picofarad equivalents, you would see the reasoning in marking since different values require more space. Obviously it would take less space to print "1" or 1pF than 0.000001mF or 820pF instead of 0.00082mF. As the decimal equivalent approaches unity, you may find either 0.001mF or 1000pF used, but the division line exists here because 0.01mF is easier than its 10,000pF equivalent.

4.1-5   Capacitor Testing

Two vitally important characteristics of capacitors provide common sense type troubleshooting methods. The characteristics common to DC type capacitors are: blocking action offered to DC voltages and passing action offered to AC voltages.

The DC blocking characteristic is possible because once the capacitor charges to the applied potential it remains at that potential. The AC passing characteristic is a result of capacitor discharge and recharge when varying signals affect circuitry operation.

4.1-5A   OHMMETER TESTING

Ohmmeter circuitry design, consisting of a dc voltage source and voltage divider (resistive) circuitry, enables us to use the time constant or charge time—discharge time meter readings as a primitive capacitor test. If a capacitor, one lead removed from the circuit with power off, is discharged and then checked with an ohmmeter, a meter needle rise and fall (kick) should be observed when charging and discharging the capacitor via ohmmeter battery supply voltage. To do this, connect the ohmmeter leads across the capacitor to charge, then reverse the leads to discharge. A small rise-fall meter needle indication should result for small-valued capacitors. Electrolytic capacitors will provide greatest meter deflection due to their large capacitive value.

As a general rule, since the larger the capacity, the larger the kick, smaller valued capacitors from 0.1mF into picofarads will require higher ohms multiplier settings while higher capacitances would require lower multiplier settings because of charge-discharge times and meter deflection responses. It is normal when electrolytic capacitors are checked for the needle to peg due to discharge (lead reversal), but do not allow it to remain pegged—drop to a smaller multiplier (another time constant) to avoid meter damage. Normally a capacitor is defective when it does not show a rise-fall indication or indicates a constant resistance (providing the multiplier is not in megohms, etc). A better test naturally is using a capacitor tester or by substitution.

4.1-5B   VOLTMETER TESTING

Capacitors usually fail in three general ways: 1) failure due to break in leads inside encasement, 2) failure due to dielectric breakdown causing shorting action, 3) failure due to leakage due to dielectric deterioration, a step just before shorting action. Naturally, capacitor failure symptoms cause various things to happen depending upon their use in the circuit. Voltage checks will aid when troubleshooting circuitry if your theory is up to date.

If you suspect a leaky capacitor, DC voltages will no doubt be passed through the capacitor. One way to find out is by removing one end of the capacitor from the circuit (keep the hot DC lead connected) and then connect a DC voltmeter from the free end to circuitry ground. Any DC voltage present indicates a leaky condition because capacitors are supposed to block DC voltage. Bypass capacitors going to ground can mess up voltage checks due to shorts or leaky conditions also, so to be sure which one is guilty, cut or unsolder one of its leads thereby removing any doubt.

4.1-5C   CAPACITOR SHUNTING

The best way to check a capacitor thought to be defective is by substitution with a known good one, but this requires added work and perhaps a loss in valuable time. Capacitor shunting is a method often used to verify a defective capacitor. Since capacitors connected in parallel in effect add capacitance, leaky or open capacitors can be found quicker. All capacitors used when shunting other capacitors must have the same working voltage rating or higher. It is necessary to observe and match polarity markings for all electrolytic capacitors. Capacitors of the non-electrolytic types need not be polarity matched. However, to be professional, replace the capacitor with the same value and in the same polarity connection.

4.2   CHASSIS AND INDUSTRIAL WIRING COLOR CODES

Manufacturers of electronic components and devices have for years marked their products with specific color codes which have specific meanings. The trained electronics person can readily identify facts concerning resistors, capacitors, inductors, transistors and even circuitry functions.

Industrial and commercial electronic circuitry wiring have accepted color codes which help you and me to identify circuitry functions. The unfortunate part of identifying circuits solely upon wire colors used is that of reliability because not all manufacturers use accepted colors for specific circuitry connection. The colored wiring assignments listed in this chapter usually can be relied upon, but due caution must be observed when equipment circuitry identification by colors is in doubt. To be safe, verify first and rely second.

Wiring Identification

In wiring harnesses or in places where wires cannot be easily traced from beginning to end, manufacturers identify each wire-circuit by color, or by color and number, or simply by number. The resistor color code digit assignment provides quick relationships that are frequently used. To the left of each color listed for the wire colors employed in chassis and industrial wiring assignments you will find a digit representing that particular color.

4.2-1   Commercial Radio-Television Chassis Wiring Color Code

		Digit	Color	Circuitry Function
4.2-1A	**GROUNDS**	0	Black (solid)	grounded components and returns
			Black-brown	Identified grounds i.e., filament
			Black-red	Identified grounds i.e., B—
			Black-yellow	Identified grounds i.e., cathode/emitter
			Black-green	Identified grounds ie; grid/base
4.2-1B	**HEATER/FILAMENTS**	1	Brown (solid)	Above or below ground
			Brown-red	Identified purpose i.e., rectifier
			Brown-yellow	Identified purpose
			Brown-green	Identified purpose
			Brown-white	Identified purpose
4.2-1C	**POWER SUPPLIES**	2	Red (solid)	Main source, B+
			Red-black	Intermediate source potential
			Red-yellow	Identified source potential
			Red-green	Intermediate source potential
			Red-blue	Intermediate source potential
			Red-white	Identified source
			Red-blue-yellow	Intermediate source potential
4.2-1D	**SCREEN GRIDS**	3	Orange (solid)	Positive potential

		Digit	Color	Circuitry Function
4.2-1E	CATHODES/EMITTER	4	Yellow (solid)	Above and below ground
			Yellow-red	Identified circuit i.e., output
			Yellow-green	Identified circuit i.e., oscillator
4-2.1F	CONTROL GRIDS/BASES	5	Green (solid)	Bias potentials
			Green-red	Identified element
			Green-yellow	Identified element
			Green-white	Identified element
4.2-1G	PLATE/COLLECTOR	6	Blue (solid)	Plate or collector potentials
			Blue-red	Identified element
			Blue-yellow	Identified element
4.2-1H	MISCELLANEOUS	7	Violet (solid)	Biases, returns
4.2-1I	AC POWER SOURCE	8	Gray (solid)	AC power potentials
4.2-1J	BIAS SUPPLY SOURCE	9	White (solid)	Main source
			White-black	Alternate/offground connection
			White-brown	Intermediate AVC bias
			White-red	Below ground, maximum value
			White-orange	Intermediate fixed value
			White-yellow	Intermediate fixed value
			White-green	Preferred AVC bias
			White-blue	Internal antenna or connection to

4.2-2 Industrial Wiring Color Codes

Color	Abbreviation	Common Circuit Function
Black	Bk	Line not at ground potential, load, line voltage control
Red	R	Alternating current control, not at ground potential
Yellow	Y	Interlock panel control
Green	Gn	Safety grounding conductor
Blue	Bl	Direct Current Control
White	W	Grounded Neutral Conductor

4.3   SEMICONDUCTOR DEVICES

Perhaps someday, standards applied to semiconductor related components will be as clear, concise, and reliable from manufacturer to manufacturer as are color codes assigned to axial lead resistors. The agencies who, out of necessity, have contributed to the standards manufacturers follow and we rely upon have great difficulty with semiconductor standardizations being assigned and adapted. Part of this problem is due to the rapid increase in the number of different semiconductor devices while another part is the difficulty in convincing manufacturers that their particular design is poor and that the agency type, size, shape, style, or number is best.

This section of Chapter 4 will share with you those semiconductor facts which seem to be commonplace in manufacturer practice although not necessarily "approved" as standards for that particular device.

4.3-1   Packaging Standards

Prefixes such as "TO," standing for Transistor Outline, were meant to define transistors having more than two leads; "DO" meaning Diode Outline, defining diodes having only two terminals or leads; "MO" meaning, I think, Metal Oxide for devices having five or more leads; and the "SP" prefix which I believe stands for Special Purpose. These prefixes are commonly used to describe semiconductor devices, but not exactly as planned by the standards agencies. The "TO" prefix, for example, would describe only transistors having more than two leads. This prefix would allow recognition by physical encapsulation measurements and specifications. Numbers were to be added to the "TO" prefix according to organizational standards, thereby making TO-1 the first registered and hopefully accepted size, etc. and additional numbers indicating its registered "TO" number. Diode outline "DO" and other prefixes would follow in the same manner.

Most manufacturers of semiconductor devices do use suggested lead spacing tolerances, lead diameter tolerances and in general the size-outline with their assigned prefixes, but modifications are normally included. As an end result, the "TO" prefix for transistors has application for germanium diodes, triacs, diacs, silicon controlled rectifiers, light emitting diodes, integrated circuits, plus more. We also find the "DO" prefix applied to light emitting diodes, diodes, rectifiers, and diacs. Prefix letters are substituted by style, case or number-letter combinations plus subscripts, suffixes, and abbreviations. Some of these identification type prefix substitutes refer to press fits, stud mounts, and/or heat radiators possibly used on or with semiconductor devices.

With all of these confusing facts, you can see the problem existing because one fact may not be true for another manufacturer's semiconductor device. Some of the more popular prefixed shapes or packaging design characteristics are illustrated in Figure 4-2A, B, C. The illustrations merely illustrate common similar outlines often used, terminal/lead identification, and terminology encountered regardless of packaging styles manufactured. It must be remembered that each prefix, or case number has its

TO1
TO44

TO3
TO41
TO66

TO5, TO33, TO43, TO66, TO77, TO78

TO7
TO45

TO8, TO18,
TO71, TO72
TO104

TO9

TO12

TO36

TO39

TO40

TO46
TO52

TO48, TO64
SP69, SP70
SP81, SP95

TO53

TO59, TO60
TO61, TO63

TO89

TO92
SP111
SP112

TO98
SP104

TO105
TO106

Figure 4-2 (a)

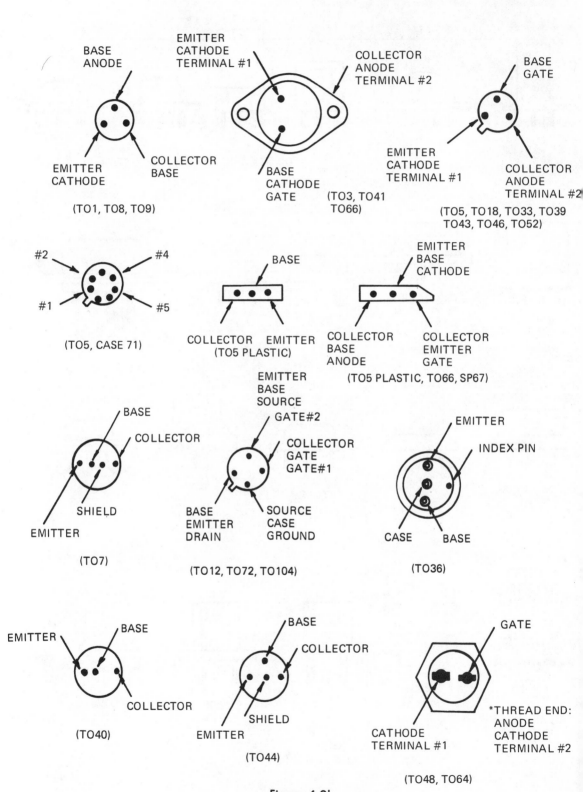

Figure 4-2b

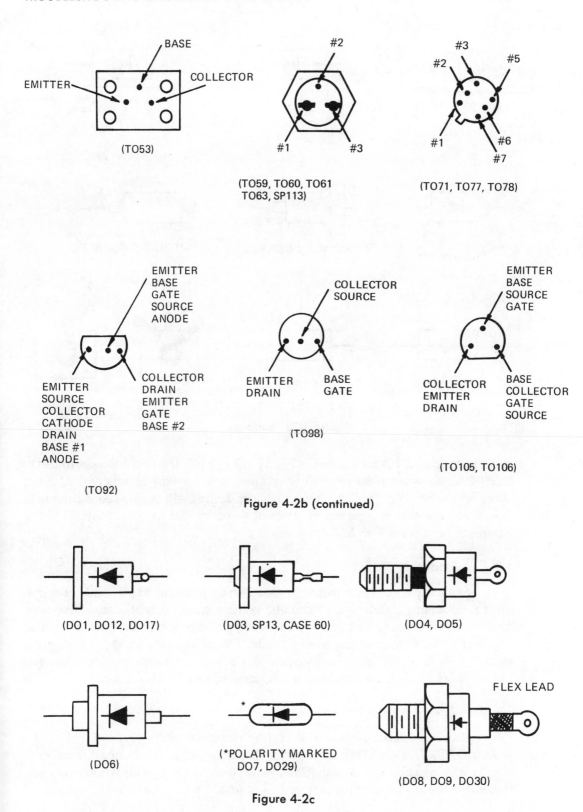

Figure 4-2b (continued)

Figure 4-2c

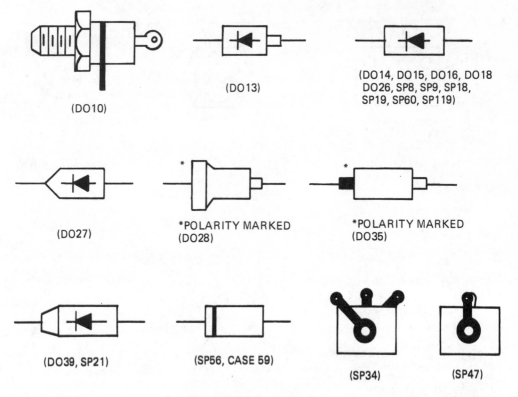

(DO10)

(DO13)

(DO14, DO15, DO16, DO18
DO26, SP8, SP9, SP18,
SP19, SP60, SP119)

(DO27)

*POLARITY MARKED
(DO28)

*POLARITY MARKED
(DO35)

(DO39, SP21)

(SP56, CASE 59)

(SP34)

(SP47)

Figure 4-2c (continued)

own dimensions and is not implied in Figure 4-2A, B, C. Individual facts pertinent to specific semiconductor devices will be outlined in the following subdivisions listed under packaging. For simplicity, semiconductor devices will be grouped according to typical names rather than packaging styles or shapes so that specifics pertinent to that group can be covered with less confusion.

4.3-2   Diodes

Semiconductors in the particular diode family grouping may be made of copper sulfide, selenium, germanium, silicon and perhaps other materials, depending upon their electronic circuitry function. Their function might be as a zener, rectifier, current regulator or reference-steering diode. Diodes, regardless of purpose, are normally identified by an "IN" letter combination plus a series of numbers. Since only two terminals exist, the cathode and the anode, identification is made easier.

4.3-2A   ZENER DIODES

Packaging zener diodes could best be illustrated by the prefixed outlines for DO4, DO5, DO7, DO13, TO3, and TO92, shown in Section 4.3-1. Zener diodes have a wattage rating ranging from about 250 milliwatts to 50 watts. Normal zener voltages, sometimes marked on their packaging, ranges from 1 volt to hundreds of volts.

The cathode side of the zener dioded is polarity marked when packaged in the DO7 by a band or color closest to one lead, while the DO 4 and DO5 may have either end marked as the cathode. When confronted with these two types, rely on the diode symbol stamped on their case when determining the cathode end. The TO3 style usually uses the anode lead and case, while the remaining lead is the cathode.

In some cases, suffixes added to the end of the diode, i.e., IN5885"A" denote tolerances. In most cases if no suffix is used, a ± 20 percent tolerance is indicated: "A" suffix means ± 10 percent; "B" means ± 5 percent; "C" means ± 2 percent; and "D" suffix means ± 1 percent. Zener diodes having a suffix resembling R, RA, RB or something similar may refer to reverse polarity types and should be verified before using.

4.3-2B   ZENER REFERENCE DIODES

Zener reference diodes normally use voltages from about 5 volts to 20 volts while temperature compensated. Zener reference diodes have voltage ranges from about 12 volts to hundreds of volts. The previous data also applies to these reference type diodes.

4.3-2C   CURRENT REGULATOR DIODES

Current regulator diodes, normally packaged in the usual DO7 glass case style, cover current ranges from about 200 microamperes to 5 milliamperes. Often they are connected in series to extend possible limiting voltages and in parallel to increase the current capabilities needed. Polarity marking and suffix data previously mentioned hold true for current regulators.

Diode testing can be accomplished using an ohmmeter, since high resistance should result when reversed-biased (negative on the anode and positive on the cathode), and low resistance is measured when the cathode is negative and the anode is positive. The greater the difference in forward and reverse bias when ohmmeter testing, the better the diode. Normally a 10 to 1 ratio, when on RX1000, is acceptable. Make sure of your meter's battery-ohmmeter polarity because in some meters the black lead does not indicate the negative side of the battery used for ohm readings.

When testing zener diodes via ohmmeter methods, make sure the zener is rated above the ohmmeter battery potential.

4.3-3   Field Effect Transistors

Conventional transistors of the "NPN" or "PNP" variety are called bipolar transistors because this variety depends on electron charge carriers and hole charge carriers. Field effect transistors, a unipolar semiconductor device, by comparison have one charge carrier which may be either positive or negative. The basic FET has three terminals, a gate, drain, and source. The gate, acting like a vacuum tube control grid, affects the current flow between source and drain terminals or vice versa. There are, however, FETS with more than one gate, making them ideal for special circuitry functions.

The field effect transistor types normally are referred to by general gate structure; junction field effect transistor (JFET), and metal oxide semiconductors, along with insulated gate field effect transistors (MOSFET and IGFET). Some of the normal packaging prefixed outlines are TO18, TO72, TO76, and TO92. Refer to Section 4-3-1 for illustrations.

FETs normally are listed by a "2N" number-letter combination, plus additional numbers when descriptions are made. These semiconductors may use "N" or "P" channel designations, meaning that the gate bias is different depending upon channel type.

Most JFET applications use fixed or a form of self bias to insure the gate will *not* be forward-biased by a signal. For example, an "N" channel JFET has a negative bias from gate to source, while a "P" channel uses positive bias. MOSFETS, on the other hand, operate with both bias voltages. MOSFETS are also described as being depletion types or enhancement types. Depletion types allow current to flow with zero bias—additional amounts of bias will reduce current flow. Enhancement types don't allow much current flow until the bias is such as to turn the device on.

Ohmmeter indications, preferably VTVM types, should tell, via resistance readings, source and drain terminals of a FET since their resistances between each other will be the same regardless of lead switching. The remaining lead (out of three) will be the gate.

Ohmmeter readings for JFET testing are good for gate identification since diode type resistance readings will appear, high resistance or low resistance, depending on lead polarity when one of the three leads happens to be the gate.

4.3-4 Germanium Power Transistors

Wattages ranging from about 100 milliwatts to hundreds of watts are obtainable using germanium power transistors. The packaging styles illustrated in Section 4-3-1, vary from TO-3, TO-36, TO-41, TO-66 to various other styles and shapes. Normally they are grouped according to specific purposes, as drivers, outputs, currents, high voltage, etc. Most manufacturers identify them by assigning "2N" prefixes plus numbers. The collector's current maximum value (IC max) is an important factor used when selecting germanium power transistors. Manufacturers normally give this pertinent information when describing their particular transistor. Collector currents from about 1 ampere to 50 and 60 amperes are normal, but currents of 100, 150 and higher are possible when power packages are assembled.

4.3-5 Integrated Circuits

Another name associated with integrated circuits is called micro-electronics which encompasses all types of very tiny circuits or circuitry systems. Integrated circuits (IC) might be grouped in three general package-style related terms. One group is the *flat-pack* style, the second group called *dual-in-line* style, and the third includes *integrated chip* and *beam lead* styles.

The flat package comes in a variety of sizes each having differing number of leads or terminals. Normal flat packaging sizes are about ¼ × ¼ having 10 and 14 leads; ¼ × ⅜ with 16 and 24 leads; and ⅜ × ⅝ with 24 leads. The "flat pack" illustrated in Figure 4-3 indicates a reference marking system normally used to point out terminal number one.

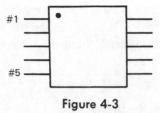

Figure 4-3

Dual in-line packages differ in style, measurements, and lead number. However, somewhat standardized sizes are available. Some of the approximate shape measurements include the $^9/_{32} × ^{25}/_{32}$ having 14 and 16 leads; the $^{33}/_{64} × ^{55}/_{64}$ with 16 leads or $^{33}/_{64} × 1^{55}/_{64}$ having 36 leads; other size variations are packaged with 8, 14, 16, 20 and 24 leads. The "dual in-line" package illustrated in Figure 4-4 indicates a marking system commonly employed, denoting the number one terminal, thereby providing standardized pin count.

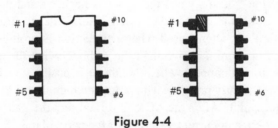

Figure 4-4

Integrated chips—beam lead microelectronic circuits—are normally mounted, when used, by a glue or paste. Their very fragile packaging sizes range from approximately $^1/_{32} × ^1/_{32}$ to about $^1/_{16} × ^3/_{32}$. As in the previous two integrated circuitry packages, a marking system is employed to identify pin or lead references. Figure 4-5 illustrates a typical chip or beam lead package.

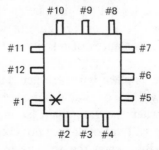

Figure 4-5

As with most other semiconductor packaging situations, TO-5 outlines and others are common. Integrated circuits manufactured by some companies can be identified by their letter or number sequence used before the 4 or 5 digit number used to describe the integrated circuit purpose or function. Suffixes are also used to describe the packaging style or specific purpose. Dual in-line ceramic packages use the letter "L" suffix; dual in-line plastic packages are identified by the letter "P"; flat packaging is denoted when the letter "F" appears as the suffix; the letter "G" indicates metal can packaging; and if the integrated circuit is a driver, the letter "A" as a suffix is usually found.

4.3-6   Silicon Controlled Rectifiers

Silicon-controlled rectifiers (SCR), are normally rated in "on state" maximum forward, rms current values ranging from 0.5 amperes to 80 or 100 amperes. The packaging styles of TO-5, TO-18, TO-39, TO-48, TO-64, TO-66, TO-83, TO-92 and TO-94 merely confuse those of us who have been around because they look like three-lead transistor types. SCR leads are identified as being a cathode, an anode, and a gate. The gate-cathode resistance will be about the same value even when switching the ohmmeter leads. An infinite resistance (or nearly) will be observed when resistances are taken between anode and gate or anode and cathode.

"Diode action" occurs when the anode is negative and the cathode is positive. The SCR will not or should not conduct when in this state, regardless of gate voltage.

"Gate action" is a term used when conduction is prohibited because the gate voltage is equal to or more negative than the cathode when the anode is positive.

Full SCR turn-on happens when the anode is positive; cathode is negative; and when the gate is sufficiently positive. The point at which the SCR is turned on is called the trigger level or point. After the SCR is turned on, the gate potential has lost its effect; the anode voltage or current specifics of the particular SCR now control conduction. If the voltage/current drops below a certain value, the SCR is turned off and requires gate potential aid.

4.3-7   Triacs

Triacs are similar to SCR semiconductors in several ways. A triac has three leads; a gate and two anodes which are referred to often as being main terminal #1 and main terminal #2. Packaging styles are similar, being TO-5, TO-48, TO-64, etc. Triacs also are rated according to rms forward current values except from about 1 ampere to 40 and 50 amperes. Resistances of infinity or nearly infinity should result between either anode to gate.

Triac "diode action" takes place when the gate has no turn-on potential with one anode negative. The triac will not or should not conduct.

"Gate action" occurs when the gate again has no turn-on potential with an anode positive. Again, the triac will not or should not conduct.

Full conduction is possible when the triac is in one of the following two situations: if one of the two anodes is positive and the gate is positive enough (past its particular trigger point), or when one of the two anodes is negative and the gate potential is sufficiently negative. In any case, once the triac is turned on, any change in gate voltages (positive or negative) has no effect. The triac turns off when its voltage or current drops below the particular holding point value and requires gate action to turn it back on.

4.4   FUSES AND CIRCUIT BREAKERS

An inexpensive insurance investment for any electronic device is a properly selected and installed fuse or circuit breaker. Like most insurance policies, there is one just right for your particular need. Fuses are described according to their relationship between the current value flowing through them and the time it takes for its interrupt function to occur. The common terminology used to describe fuse types is (1) "fast acting," "quick acting," "high speed" or "instrument"; (2) "standard," "normal lag," "normal" or "medium lag"; (3) "time delay," "time lag," "slow acting," or "slow blow." Each type, obtainable in differing current carrying capacities, will protect electronic circuitry if the fuse interrupt-time element is fast enough.

4.4-1   Voltage-Current-Fusing Ratings

All fuses have specific voltage, current, and fusing ratings. All of these ratings apply to slow-acting, medium-acting and fast-acting fuses, regardless of physical size.

The voltage rating marked on the fuse is an Underwriter's Laboratory guarantee for fire risk. What it says is that the fuse will safely open without arcing or exploding in a short circuit situation (10,000 amps) when the voltage is equal to or less than the rated voltages. Do not, under any circumstances, use a fuse rated at lower voltages than the voltage applied to the fused circuitry, regardless of its amperage rating. The fuse, however, may be used at any voltage less than its voltage rating without affecting its designed fusing characteristic.

The current ratings marked on fuses indicate the current load standardization testing value. Fast-acting fuses are designed to carry 100% of their current rating, but will blow quite rapidly when their current load is exceeded by a small percentage. Medium-acting fuses generally are designed to carry 110% of their rated current for a minimum period of four hours or at 135% of their rated current for periods less than one hour, or at 200% of their rating current for a maximum of 30 seconds. Slow-acting fuses, on the other hand, are designed to carry 110% of their rated current for a minimum period of 4 hours, but if 135% overload current is reached, will open within one hour. If the slow-acting fuse has a current value flowing through it equal to 200% of its rated value, it will stop current flow sometime within a 5-second to 2-minute delay period.

4.4-2   Fuse Standardizations

It seems that technology growth has pushed aside logical standardizations for fuses by physical size or number letter combination. Efforts to prevent overfusing by adding to fuse length or differing diameter sizes, etc., further add to confusion. Some of the more popular fuse manufacturers identify fuses by letter, a number-letter combination, or a purely number identification plus fuse values. Physical sizes, amperage range and voltage range are illustrated in the following data:

FAST-ACTING

Common Identification	Amperage	Voltage	Physical Size (in)
3AG, 312	1/100-10A	250V, 125V	¼ × 1¼
4AG, 412	1/16-8A	250V, 125V	$^9/_{32}$ × 1¼
5AG, 512	1-8A	250V, 125V	$^{13}/_{32}$ × 1½
7AG, AGW, 303	1-30A	32V	¼ × ⅞
8AG, MJV, 361	1/500-30A	250V, 125V	1 × ¼
Type "F," 212	1/10-6A	250V	0.197 × 0.787

MEDIUM-ACTING

	Common Identification	Amperage	Voltage	Physical Size (in)
	1AG, AGA, 301	1-30A	32V	¼ × ⅝ (⅞)
	3AG, 311	4-40A	32V	¼ × 1¼
(1)	3AG, GJV, 318	1/100-30A	250V, 125V	0.275 × 1.275
(2)	3AG, AGC, 3AB, 334	3/4-20A	125V, 32V	¼ × 1¼
	4AG, AGS, 411	10-40A	32V	$^9/_{32}$ × 1¼
	5AG, AGU, 511	10-60A	32V	$^{13}/_{32}$ × 1½
	8AG, AGX, 362	1/8-30A	250V, 125V, 32V	¼ × 1
	3AB, ABC, 314	1/8-30A	250V, 125V	¼ × 1¼
	4AB, ABS, 414	1-40A	250V, 125V, 32V	$^9/_{32}$ × 1¼
	5AB, ABU, 514	1-30A	250V, 125V	$^{13}/_{32}$ × 1½
	SFE, 307	4A-30A	32v	¼ dia, lengths ⅝-1$^7/_{16}$
	Type M, 211	1/50-5A	250V	0.197 × 0.787
(3)	Type C, 332	1/32-10A	250V, 125V	

SLOW-ACTING

	Common Identification	Amperage	Voltage	Physical Size (In)
	3AG, MDL, 313	1/100-30A	125V, 32V	¼ × 1¼
	3AG, MDV, 315	1/100-30A	125V	0.275 × 1.275
	4AG, MDM, 413	1/16-30A	125V, 32V	$^9/_{32}$ × 1½
	5AG, MDR, 513	1/2-30A	250V, 125V, 32V	$^{13}/_{32}$ × 1½
	3AB, 323	1/100-30A	250V, 125V, 32V	¼ × 1¼
	4AB, 423	1/16-30A	250V, 125V, 32V	$^9/_{32}$ × 1¼
	5AB, 523	1/10-30A	250V, 125V, 32V	$^{13}/_{32}$ × 1½
	Type "T," 213	1/10-6A	250V	0.197 × 0.787
(3)	Type "N," 333	1/100-7A	125V	

SUBMINIATURE

Common Identification	Amperage	Voltage	Physical Size (In)
Picofuse	1/16-15A	125V, 32V	$^3/_{32} \times {}^9/_{32}$
Microfuse	1/500-5A	125V	0.250×0.348

1. Pigtail types
2. Indicating types
3. Differing length and tab width dimensions

4.4-3 Circuit Breakers

Electronic circuitry having manual or automatic resetting circuit breakers enables you and me to save time since the device need not be taken apart simply to re-fuse the circuit. Circuit breakers have definite current carrying capabilities and often are described by normal current rating, break current and hold current ratings. The break current value describes the amount of current the breaker will trip at, thereby protecting the circuit of greater current values. The hold current value indicates the minimum value of current allowable for that particular breaker. Any value under the hold current amount will not allow the circuit breaker to reset as designed.

Typical manual reset circuit breakers, normally rated at 125 volts, are listed next, while the second listing typical automatic resetting types, rated normally at 6 volts to 24 volts, are indicated.

MANUAL RESET

Current Rating	Break Current	Hold Current
0.650A	0.86A	0.49A
0.80A	1.05A	0.60A
1.00A	1A-1.20A	0.65A
1.25A	1.63A	0.93A
1.50A	1.5A-1.75A	1.00A
1.75A	2.10A	1.20A
2.00A	2A-2.60A	1.25A-1.50A
2.25A	2.90A	1.65A
2.50A	2.5A	1.60A
2.75A	3.30A	1.90A
3.00A	3A-3.67A	1.90A-2.10A
3.25A	3.85A	2.20A
3.50A	3.5A-4.03A	2.20A-2.30A
3.75A	4.20A	2.40A
4.00A	4A-4.40A	2.50A
4.50A	4.5A-5.25A	3.00A
5.00A	5A-5.70A	3.25A
5.50A	5.50A	3.60A
6.00A	6A-6.82A	3.9A
7.00A	7A-7.25A	4.14A

AUTOMATIC RESET

Typical Current Ratings

     5A, 6A, 8A, 10A, 12A, 15A, 18A, 20A, 25A, 30A, 35A, 40A, 45A, 50A

Breaking Time

     100% load; no breaking—arcing
     125% load; open within 1 hour
     200% load; open within 30 seconds, resets within 10 seconds

4.5   RESISTORS

All resistors, whether insulated or non-insulated, composition, metal glaze, glass-tin-oxide, wirewound, adjustable or variable, fusable, fixed film, linear or non-linear taper, have specifications which must be met by manufacturers before standardization agencies will approve their use. These standards provide you and me with reliable facts regarding identification by codes, sizes, and shapes. They also insure chances of greater interchangeability since uniform methods and specifications are used when testing before selling. Often standards are referred to on resistor packaging and/or on data sheets. Two of the abbreviations often used are EIA (Electronic Industries Association) and MIL (Military). There are several more agencies involved and page after page of specifications which must be met to rate special letter-number coding.

4.5-1   Composition/Carbon Resistors

4.5-1A   RESISTIVE ELEMENT

Carbon or graphite mixture formed into ceramic type core.

4.5-1B   RECOGNITION FEATURES

Standard EIA-MIL color coding. If axial lead *insulated,* the body color is *usually tan*, but other colors, except for black, are used. If axial lead non-insulated, the body color is *usually* black.

4.5-1C   WATTAGE RATINGS

Normal wattage is ⅛, ¼, ½, 1, and 2 watts. The "MIL" letter-number assigned ¼ watt is RCR07, for ½ watt; RCR20, one watt is described by RCR32, and the two-watt resistor by RCR42. The physical size determines wattage assignments in this grouping. The suggested voltage rating for a wattage of ⅛ watts is 150 volts, ¼ watt is 250 volts, ½ watt uses 350 volts, 1-watt maximum voltage is 500 volts as is the 2-watt rating.

4.5-1D   RESISTIVE VALUES

Resistive values from 2.7 ohms to 22 million ohms having a five-or ten-percent tolerance are common. Twenty-percent tolerance resistors range normally from 0.1 ohms to 22 million ohms.

4.5-2   Metal-Glaze/Glass Tin Oxide Resistors

4.5-2A   RESISTIVE ELEMENT

Layer of film of glass and metal fused into a crystaline-ceramic type core or fused tin oxide deposits to glass or ceramic surface in spiral paths.

4.5-2B   RECOGNITION FEATURES

Standard color coding, tin-electroplated copper leads, solvent resistant molded body; resistive values may also be observed on resistor body.

4.5-2C   WATTAGE RATINGS

Normal rating is ¼ and ½ watts.

4.5-2D   RESISTIVE VALUES

Resistors ranging from about 50 ohms to about 150 thousand ohms are common, having two-percent and five-percent tolerances.

4.5-2E   ADVANTAGES

Closer tolerance due to manfuacturing processes. More stability throughout operating temperature ranges.

4.5-3   Metal Film Resistors

4.5-3A   RESISTIVE ELEMENT

Thin layer of resistive material deposited on insulated cores makes up low-resistance values; equivalents to wire-wound resistors are possible by using spiral patterns; coarse pattern for intermediate resistances and close spirals for high resistance.

4.5-3B   RECOGNITION FEATURES

Smaller in physical size than carbon types, may be color coded or stamped. Body types include molded and conformally coated. Letter "M" means molded, while "C" means conformally coated.

4.5-3C   WATTAGE RATINGS

Vary depending upon temperature, resistance, and standards agency or manufacturer. What the manufacturer calls ¼ watt, "MIL," standards say perhaps $1/30$ watt. Manufacturing labels for wattage range from ⅛, ¼, ½, ¾, 1, 1½, and 2 watts.

4.5-3D   RESISTIVE VALUES

Resistance values from 10 ohms to 1.5 million ohms are common. Maximum voltages are also given for specific resistances at manufacturer's suggested wattage. Resistors rated at ⅛ and ¼ watt have 200 volts maximum rating; ½ to ¾ watt have 250 volts maximum; 1 to 1½ watts have 350 volts maximum, and 2 watts have up to 700 volts maximum ratings. Make certain of wattage and/or voltage limitations for your use.

4.5-3E   ADVANTAGES

More stability than encountered in composition types and are smaller physically.

4.5-4   Power-Wire Wound Resistors

4.5-4A   RESISTIVE ELEMENT

Wire wound resistors normally have low resistances and are capable of carrying high amounts of current. There are many different techniques used in manufacturing and, therefore, physical size is of little importance. Normally, nickel-chromium or copper-nickel alloy wound around a tubular ceramic material is employed.

4.5-4B   RECOGNITION FEATURES

Sometimes they are color coded. If this is the case for a fixed wire-wound resistor, the first color band is about twice the width of the other color bands. Often the case is rectangular or circular made of ceramic materials, but metal heat-sink types are also popular. Resistive values and wattage ratings are usually printed on its case.

4.5-4C   WATTAGE RATINGS

Ratings from ½ watt to the hundreds are popular. Their wattage rating can be extended somewhat if their heat can be dissipated by their positioning or mounting.

4.5-4D   RESISTIVE VALUES

Resistance values from about ½ ohm to hundreds of thousands of ohms are manufactured. However each specific wattage group may have its own preferred or standard resistive values.

Potentiometers, fusable, temperature and voltage sensitive resistors also have their standards, but they seem somewhat less secretive and therefore will not be covered. The remaining data will be dedicated to color marking relationships of the axial and radial lead resistors, which compose the two groups of fixed type resistors. For clarity's sake, axial lead resistors have one wire lead protruding from both ends of the resistor, while the radial lead resistor has its leads wrapped around the resistor body.

4.5-5   Fixed Resistor—Preferred Resistive Values

Fixed composition resistors normally have three general tolerance groupings which seem manufacturer established and agency approved. Thre three groupings of ± 5% ± 10%, and ± 20% tolerances assigned to specific quality controlled resistors eliminate manufacture of resistors in a sequential resistive progression. This is possible since resistance of the ± 20% nature would duplicate the 10% resistive values, and in like manner, the 10% values would overlap into the 5% values.

To prevent resistive value duplication, a system called ";preferred value" controls resistor value production. The preferred value system uses a specific multiplier which, when multiplied times the beginning resistive value within that tolerance group, equals or closely equals the next preferred resistive value. After multiplication by the given multiplier, the end result value is a rounded-off, two-significant figures digit. For example, one multiplier for a 5% resistor is 1.58 and the beginning resistive value is 10 ohms, so the next preferred value is: 10×1.58, equaling 15.8 ohms or rounded off equals 16 ohms. This data is outlined within the following chart.

TOLERANCES	5%		10%		20%
Two U.S. Standard Multipliers *Z17.1 **C83.2	*1.58	**1.10	*1.26	**1.21	1.46
	10	10	10	10	10
	—	11	—	—	—
	—	12	12	12	—
NOTE:	—	13	—	—	—
	—	15	—	15	15
(1) For resistances under 100 ohms, multiply	16	16	16	—	—
by one.	—	18	—	18	—
	—	20	20	—	—
(2) For resistances under 1000 ohms, multi-	—	22	—	22	22
ply by ten. IE: 160 ohms may be bought	—	24	—	—	—
having 5- and 10-percent tolerances.	25	—	25	—	—
	—	27	—	27	—
(3) For resistances under 10,000 ohms, mul-	—	30	—	—	—
tiply by one hundred. IE: 2,200 ohms can be	—	—	32	—	—
bought having 5-, 10-, and 20-percent toler-	—	33	—	33	33
ances.	—	36	—	—	—
	—	39	—	39	—
(4) For resistances under 100,000 ohms,	40	—	40	—	—
multiply by one thousand. IE: 62,000 ohms	—	43	—	—	—
is available only in a 5-percent resistor.	—	47	—	47	47

TOLERANCES		5%		10%		20%
Two U.S. Standard Multipliers *Z17.1 **C83.2		*1.58	**1.10	*1.26	**1.21	1.46
(5) For resistances greater, multiply by 10k, 100k, etc.	—	—	50	—	—	
	—	51	—	—	—	
	—	56	—	56	—	
	—	62	—	—	—	
	63	—	63	—	—	
	—	68	—	68	68	
	—	75	—	—	—	
	—	—	80	—	—	
	—	82	—	82	—	
	—	91	—	—	—	
	100	100	100	100	100	

Chart 4-1

4.5-6   Three-, Four-, and Five-Band Color-Coded Resistors

Before buying any resistor, you need to know two things: the desired wattage rating and the required resistance necessary. If you are professional, determining wattage is an easy enough task, but the value calculated does not grant much of a safety factor since resistors of the composition type change in resistive value when hot. A good safety design factor professionals use to prevent over-heating is doubling the *calculated* power rating.

Circuitry demands dictate values of resistance required due to ohms law applications, but these calculations can be quickly made as were those required in determining wattage. The next step is to select the resistor having the desired resistance, wattage rating, temperature coefficients, failure rates, and tolerances acceptable for your circuit. This is where we may have difficulty because often even we professionals forget some of the important facts concerning color-coded resistors. The following notes, figures, and charts will aid those of us who know, but have forgotten.

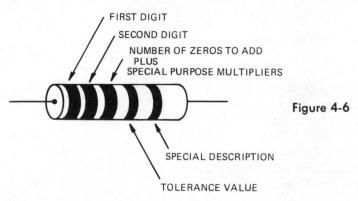

FIRST DIGIT

SECOND DIGIT

NUMBER OF ZEROS TO ADD
PLUS
SPECIAL PURPOSE MULTIPLIERS

Figure 4-6

SPECIAL DESCRIPTION

TOLERANCE VALUE

Color-coded resistors must be read correctly and therefore have standardized color positioning, color-digit or meaning assignments, and color pigmentation requirements. Although it does not make any difference as to which end is connected how in a circuit, it does make a difference at which end you begin interpreting the color codes.

4.5-6A   FIRST COLOR-CODE BAND

This is the band physically closest to the physical end of the resistor. It will not be black, silver or gold. It will be followed by at least two other close color bands. The first color band represents an assigned color digit (illustrated in Chart 4-2) which is the first number of the resistor's resistance. *Note:* If the first band appears to be double the normal color banding width, it is a wire wound resistor.

4.5-6B   SECOND COLOR-CODE BAND

This band is next to the first color. It indicates the second number of the resistor's resistance. It will not be gold or silver, but can be any of the ten colors listed in Chart 4-2.

4.5-6C   THIRD COLOR-CODE BAND

This band is used for two general purposes. The first purpose is to tell via color-digit assignment the number of zeros to add to the first and second band digits. The colors used defining the number of. zeros to add are black through white or zero-zeroes to nine zeroes. All color-coded resistors will have at least three color bands.

The second purpose is to define resistive values that are less than ten ohms and those less than one ohm. The two colors representing this second purpose are gold and silver. When silver appears in the third color band, the resistance will be under one ohm because this color means multiply the first two digits by 0.01, which also equals division by 100. If gold appears in the third color band, the resistance will be less than ten ohms because it implies multiplication of the first two digits by 0.1 or division by 10, its equivalent.

4.5-6D   FOURTH COLOR-CODE BAND

The fourth band may or may not be included on color-coded resistors. Normally this band, if present, will be separated from the first three bands by a wider space, hopefully to discourage misreading. If no fourth band exists, the tolerance indicated is 20 percent. Sometimes a black band is found where the fourth band would be; this also means 20 percent tolerance. All other colors, with the exception of green, in most cases represent a tolerance value equaling their color-digit assignment. Ten-percent resistors are identified by silver and five-percent tolerance resistors by gold when positioned in the fourth band. If green is used a guaranteed minimum value or 0 to 100% tolerance is implied.

4.5-6E   FIFTH COLOR-CODE BAND

The fifth color band may or may not be present. Normally if used, there will be four preceeding bands also. This band tells us three specific things: a failure rate, type of terminal, and test sequence.

The failure rate is for "established reliability" resistors which is actually a failure rate guarantee. These particular resistors are rated for 1000 hours of use. Although colors are used indicating a percent per 1000 hours, a letter is also assigned and might be mentioned instead of its percentage. If brown is found (letter M) the rating is 1.0; Red (letter P) means 0.1; orange (letter R) means 0.01; and yellow (letter S) means 0.001 which is the percentage given for 1000 hours.

The type of terminal, indicated for film types, uses the color white which says the terminals are solderable.

If green is found in the fifth band, this signifies that it was tested after the load cycle test.

Temperature coefficients are related in print by "MIL" characteristic letters being "K", "H", and "J". The letter "K" equals a change of $\pm$ 100 parts per million per degree centigrade (ppm/°c). "H" means $\pm$ 50ppm/°c and "J" means 25ppm/°c.

Color	1st color digit value	2nd color digit value	3rd color # o's	4th color tolerance	5th color special
Black	0	0	0	20%	—
Brown	1	1	1	1%	1.0/1khr (M)*
Red	2	2	2	2%	0.1/1khr (P)*
Orange	3	3	3	3%	0.01/1khr (R)*
Yellow	4	4	4	4%	0.001/1khr (S)*
Green	5	5	5	GMV	tested after load cycle
Blue	6	6	6	6%	—
Violet	7	7	7	7%	—
Gray	8	8	8	8%	—
White	9	9	9	9%	solderable terminals**
Gold	—	—	X0.1	5%	—
Silver	—	—	X0.01	10%	—
No color	—	—	—	20%	—

*Reliable types only — **Film types
GMV—Guaranteed Minimum Value

Chart 4-2

4.6   TRANSFORMERS

Transformers, being AC devices, have the ability to provide circuitry isolation, signal coupling, impedance matching, and voltage-current step-up or step-down action. Data and standards of importance concerning audio, IF, and power transformers will be outlined in this section of Chapter 4. The standards for color coding wires are suggested and may or may not be manufactured in accordance with them. When doubt arises, take a continuity check between the wires, pairing or matching primary and secondary leads. Center tapped leads will require closer resistance checks between the other leads.

4.6-1 Audio Output

The normal frequency of operation is between 20 hertz to 20k hertz. This particular transformer is rated in terms of its primary and secondary winding impedances, plus the wattage rating in watts. The wattage rating describes the maximum power that the secondary can deliver.

When selecting an audio output transformer, three things of importance should be done: (1) match the primary impedance closely to the impedance offered by the circuitry; (2) match the secondary impedance closely to that impedance offered by the speaker; (3) match the transformer power rating with the speaker wattage rating. The recognized audio transformer lead color coding is illustrated in Figure 4-7.

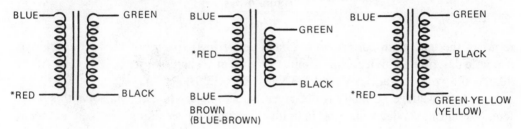

*This lead connects to power source

Figure 4-7

4.6-2 Intermediate Frequency

This transformer is designed specifically for operation at a particular frequency. The IF frequencies for standard 540 to 1600k hertz broadcast are 455, 260, and 262.5k hertz; the latter being the vehicular IF assignment. Television sound IF is 41.25 MHz while television picture IF is 45.75 MHz. When VHF is used, a 10.7 MHz IF is required while for very-ultra-and super-high frequency equipment, use IF combinations of 30, 60, and 100 MHz.

What all this means is that the IF frequency knowledge is required when replacing these transformers because specific capacitors might be used in parallel with coil windings, thereby producing the desired frequency response. This type of transformer can be checked by verifying schematic resistances of the coils if listed. Adjustments with alignment tools often are possible through top and bottom alignment holes. The repositioning of the tuning slugs peaks the frequency when correct impedance matching is obtained for that frequency. The intermediate frequency transformer lead colors are illustrated in Figure 4-8.

4.6-3 Power or Rectifier

Power or rectifier transformers have an operating frequency of 25, 50, 60, or 400 hertz, depending upon the line frequency of the alternating source. This transformer is usually larger physically than audio or IF transformers. The ratings are in

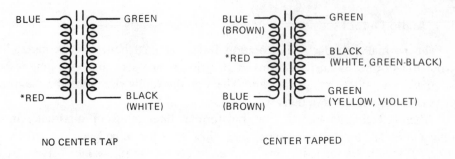

*This lead connects to the power source.

Figure 4-8

terms of the maximum current and voltage delivering capacity of the secondary. An example describing a transformer using 120 volts at 60 hertz might be 520 volts AC at 90mA, 6.3 volts AC at 3.5A, and 5.3 volts AC at 2.5A.

Total wattage normally is not given but simple calculation can solve this problem. We must first determine the individual winding power rating and then add them together. The following results, when using the previous transformer winding data.

$$
\begin{array}{llll}
520v @ 90mA & 520\times0.09 & = 46.80 \text{ watts} & P = IE \\
6.3v @ 3.5A & 6.3\times3.5 & = 22.05 \text{ watts} & \\
5.3 @ 2.5A & 5.3\times2.5 & = \underline{13.25 \text{ watts}} & \\
& & 82.10 \text{ TOTAL WATTAGE} &
\end{array}
$$

The accepted power transformer winding lead color assignments are shown in Figure 4-9.

4.7   VACUUM TUBES

Wisdom required for troubleshooting any electronic device is developed through applied knowledge of three circuits and five components. The three circuits are: (1) series; (2) parallel; and (3) series-parallel. The five components are: (1) resistors; (2) capacitors; (3) inductors (coils, transformers, etc); (4) solid state devices; and (5) vacuum tubes.

Electronic circuitry having vacuum tubes as its active device may seem easier to work on because you know they must have filament voltage, a negative cathode and a positive plate voltage before they will work as designed. Other plus factors might be that they are easily removed from the circuit for testing or replacement and have uniform, standardized structure-purpose identification traits which are adhered to regardless of manufacturer. The valuable data found within this section of Chapter 4 will contribute to your working knowledge facts which, when applied, will develop troubleshooting wisdom.

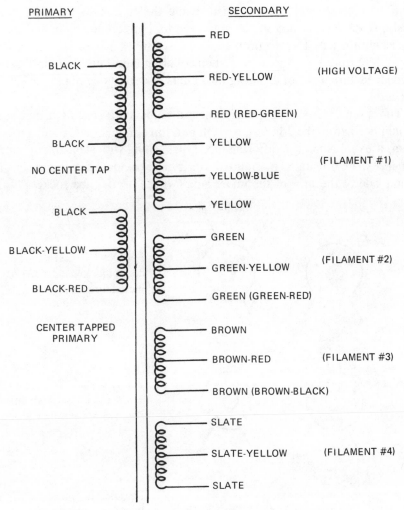

PRIMARY

SECONDARY

BLACK

NO CENTER TAP

BLACK

BLACK

BLACK-YELLOW

BLACK-RED

CENTER TAPPED
PRIMARY

RED

RED-YELLOW   (HIGH VOLTAGE)

RED (RED-GREEN)

YELLOW

YELLOW-BLUE   (FILAMENT #1)

YELLOW

GREEN

GREEN-YELLOW   (FILAMENT #2)

GREEN (GREEN-RED)

BROWN

BROWN-RED   (FILAMENT #3)

BROWN (BROWN-BLACK)

SLATE

SLATE-YELLOW   (FILAMENT #4)

SLATE

Figure 4-9

4.7-1   Pin Numbering and Physical Location

Vacuum tube envelopes may be categorized as being either metal or glass. If metal, a ceramic twelvar base (nuvistor type) or small wafer 8 pin (octal type) are common. Glass envelope types are referred to as being: small button 7 pin or 9 pin miniature; small button 9 pin (novar or noval); large button 9 pin (neonoval); small button 12 pin (duodecar); and the 8 pin (octal). The button size terminology refers to the glass indentation (button size) found on the pin side of glass tubes not having base protective-keyed envelope mountings.

Normally, the number of pins extending from the vacuum tube corresponds with the number of electrodes required. Two exceptions are some octal base tubes and some picture tubes. For example, if an octal base tube used as an indirectly heated full wave rectifier requires two plate electrodes, two filament electrodes and one cathode electrode, a total of 5 electrodes exists while the tube base normally has 8 pins. Obviously, since the tube and socket are keyed for correct pin alignment, the unused 3 pins can be eliminated. This means that octal based tubes may or may not have all eight pins protruding from the bottom, but when counting them, missing pin locations must be included. Picture tubes of differing pin counts also exist due to pin-socket alignment techniques.

Tubes not having key type alignment designs use a spacing difference separating pin number 1 from the 7th, 9th, or 12th pin. Pin elimination for the non-keyed base is seldom, if ever, used. The differing tube types, along with standardized pin numbering sequences, are illustrated in Figure 4-10. It must be noted that this relates to the pin protrusion side of the tube or the under side (wired) of the tube socket.

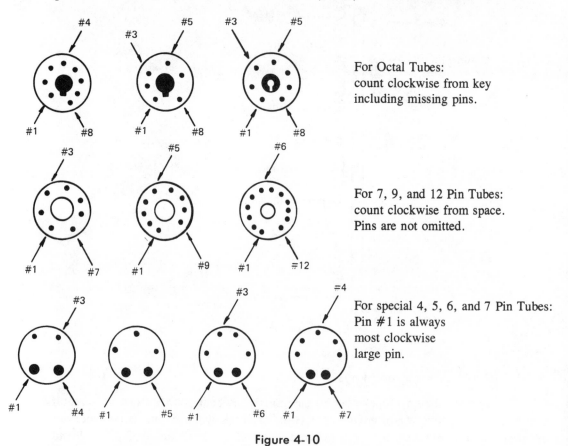

For Octal Tubes:
count clockwise from key
including missing pins.

For 7, 9, and 12 Pin Tubes:
count clockwise from space.
Pins are not omitted.

For special 4, 5, 6, and 7 Pin Tubes:
Pin #1 is always
most clockwise
large pin.

Figure 4-10

4.7-2   Tube Number-Letter Arrangement Meanings

Hundreds of thousands of vacuum tubes, differing in type, size, shape, and purpose, presently exist. Some of these tubes have one specific purpose such as diodes, triodes, tetrodes, pentodes, hexodes, or heptodes; some may combine several purposes like dual diode-triode, triode-twin pentode, etc.; and some may be classified as being 7 or 9 pin miniature, 8 pin octal, or 12 pin duodecar types, but regardless of vast differences, a logical, standardized, agency-approved method for vacuum tube purpose and identification is employed. This method is illustrated by the numerical-alphabetical tube identification found on tube price listings, substitution listings, and vacuum tube manual listings. Each tube manufactured has marked, stamped, or etched on the envelope or tube base, valuable identification data.

4.7-2A   FILAMENT VOLTAGE IDENTIFICATION

The lowest number used in the standardized vacuum tube identification method begins at zero and progresses numerically by one unit into the thousands. Normally, the digits proceeding the alphabetical letters indicate filament voltages, but as in most good things, there are some exceptions. These differences are as follows:

1. Vacuum tubes having a zero − something, normally means no filament voltage is required (IE: "0A2" = 105 volt regulator, but a "00A" requires 5 volts).

2. Vacuum tubes having a one-something through 117-something, normally indicate the approximate filament voltage (IE: 6BK4 uses 6.3 volts, a 15BD11 uses 14.7 volts, while a 117Z4 uses 117 volts, etc.).

3. Vacuum tubes having only numbers being older tube or industrial coded tube types *cannot* be relied upon for filament voltage requirements, (IE: tube #19 needs 2 volts, tube #78 uses 6.3 volts, while tube #1621 requires 6.3 volts, etc.).

4.7-2B   ALPHABETICAL LETTER IDENTIFICATION

Most all vacuum tubes have a number-letter-number-letter identification grouping code. The one or two letters following the filament numbers serve as organizational helpers for manufacturers and you and me, since they separate by alphabetically arranged sequence tubes, possibly having the same set of filament numbers. Organization is possible because letters assigned begin with "A" and go through the alphabet, then a second letter is issued along with the first thereby making possible combinations of "AB," then "AC," etc. Common 6-volt filament tubes, each differing somewhat, for example, can be identified as being a 6A8G, 6AB5, 6AC6GT, 6AD7G, and so on.

The second number grouping used in tube identification comes close, if not exactly, to the number of active or used elements within the tube. A differing element count from that which is indicated via tube number identification is highly possible unless you observe the following normally true facts. The filaments, although they may require two or more pins, are considered as being one element. An element connected internally to another element is counted only as one element (i.e., suppressor grid to cathode connection).

The last letter combination grouping (tube suffix) may or may not be used since it is related to tube specifics. Tube suffix letters commonly used are as follows:

LETTER	SUFFIX MEANING
G	octal base, glass type envelope
GT	octal base, smaller than "G" type glass envelope
A, B, C, D, E, or F	"A" indicates 1st specification modification while "B" is

LETTER	SUFFIX MEANING
	the 2nd, "C" the third, etc. Later modification types *can be* substituted for earlier versions, but not the opposite.
X	Low loss base material
Y	Intermediate loss base material
/	Indicates interchangability between the tubes

4.7-3   Electrode/Element Name and Purpose

Troubleshooting can be compared to working in the dark or working with light. Working in the dark implies limited or no knowledge of how the device works, while working with light indicates the ability to apply knowledge for symptom analysis. Since the latter troubleshooting method is preferred, we must acquire knowledge concerning names, purposes, and normal operation traits before we can make educational guesses when troubleshooting. This section of Chapter 4 will shed some light on vacuum tube electrodes by using Figure 4-11 as a reference. Each numbered electrode illustrated will be examined in a building-block fashion, beginning with the diode and progressing through the pentode.

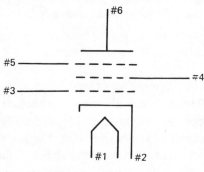

1. Filaments/heater
2. Cathode
3. Control grid
4. Screen grid
5. Supressor grid
6. Plate/anode

Figure 4-11

4.7-3A   DIODES

The diode, a two-electrode tube, requires a positive potential on its plate (#6 of Figure 4-11) plus an electron source called a filament (#1) before plate current (Ib) will exist. When only the filament (#1) and the plate (#6) are used, the diode is termed "directly heated," whereas the term "indirectly heated" is used for diodes having a cathode (#2) inserted between the filament and plate. The filaments in an indirectly heated diode merely heat the cathode thereby supplying electrons, and it is not counted as an electrode.

4.7-3B   TRIODES

The major current controlling factor of a diode, other than the correct polarity potentials on the plate and cathode, is the amount of plate potential. The higher the attracting plate potential, the more plate current will flow until a point called *saturation* occurs. The major current controlling factor in other tubes is a grid, specifically, a control grid. The triode utilizes the efficient control grid (#3 in Figure 4-11) for signal amplification purposes because small voltages applied to this third electrode can control a comparatively large amount of plate current.

4.7-3C   TETRODES

By adding a second grid to the vacuum tube illustrated in Figure 4-11, a total of four electrodes will exist; the cathode #2, the control grid #3, the screen grid #4, and the plate #6. Such a tube is called a tetrode. The extra grid, operated at a positive voltage which almost equals the plate potential, makes it possible to obtain higher amplification gains than a triode. The tetrode not only allows plate current to be practically independent of plate potentials over a certain range, but also, because of its low grid-plate capacitance, diminishes instability caused by plate-grid feedback.

4.7-3D   PENTODES

The name pentode is given to vacuum tubes having five electrodes. The fifth electrode is known as a suppressor grid, identified as #5 in Figure 4-11. This grid usually is connected to the cathode within the tube during manufacturing. As a result of the effects of the suppressor grid, power output pentodes have higher power output capabilities with small grid voltages, while in RF amplifier pentodes, a moderate-plate voltage value will provide high voltage amplification.

4.7-4   Tube Manual Terminology

Vacuum tube manual data, if understood and properly applied, can be one of the most useful tube-type circuitry troubleshooting aids presently available. When situations arise where schematics aren't readily available, reach for your tube manual because voltage and current values, pin electrode positioning, plus application advantages are listed.

In this section of Chapter 4 you will find vacuum tube specifics not only outlined, but also described, and all of them are common to vacuum tube manual data. Potential application, such as class "A" operation, class "C" amplifier, sync separator, etc., along with typical construction types like medium mu, triode, pentagrid converter, etc. will be omitted due to the vast number of differing applications and construction types presently manufactured. The data is based upon three generalized rating systems termed as being (1) average values, (2) maximum values, and (3) absolute maximum values. Each of the three rating systems concerns itself with conditions which might affect: component values, control positioning, signals, supply potentials, temperature, plus other possible important variables. In most tube manual data, average and maximum values are those suggested-normal for that particular tube.

4.7-4A   AMPERAGE

4.7-4A1.   Filaments Current—The current values for filaments, also known as heaters are for average tube operation. The first number/numbers written on the tube usually indicate the filament voltage.

   4.7-4A2.   *Plate Current*—The values listed for plate current, (Ib) usually in milliamperes, are average. Special markings, sometimes used with audio output tubes, signify the plate current was measured without a control grid signal applied. Maximum values—with a signal—are normally higher in value.

   4.7-4A3   *Screen Current*—Like the plate current values, this current is subject to the conditions of other related potential data. Extra symbols are employed in audio output tubes indicating listed current values measured without a control grid signal, etc.

4.7-4B   AMPLIFICATION FACTOR

   Mu (μ), the amplification factor for specific tubes, relates by number the ratio of small plate current changes to small grid voltage changes when plate current and other electrode voltages are kept constant.

4.7-4C   CAPACITANCE

   Normally, all capacitance values are average values; special symbols are employed to denote variations possible while measuring. These values are in picofarads (pF), formerly termed micromicrofarads (mmF).

   4.7-4C1   *Grid to Plate*—This capacitance is measured from the control grid to the plate while all other electrodes are grounded out.

   4.7-4C2   *Input*—This capacitive value relates the capacitance measured from the control grid to all electrodes except for the plate, which is connected to ground.

   4.7-4C3   *Output*—The capacitive values listed here are those measured from the plate to all electrodes except for the control grid which is grounded out.

4.7-4D   POWER

   Each vacuum tube used for amplification has its own power dissipation (wattage) curve. Plate, screen, and output wattage values differ from tube to tube and from application to application. To illustrate this, maximum plate dissipation occurs at zero signal conditions for class "A" operations, at about 65 percent of maximum signal amplitude for class "B" operations, and at a zero-signal level in a mixer or converter stage if it is at the frequency where the oscillator develops minimum bias. Plate wattage, therefore, describes a tube's safe maximum plate dissipation factor.

   The screen grid wattage listed is maximum whenever the maximum screen grid voltage is used. Special subscripts or symbols show either average, maximum, or absolute maximum wattage ratings.

   Power output wattage describes the average power output obtainable for the given operating conditions. This almost always is less circuit losses or plate dissipation minus input plate power effects. All power factors listed in the tube manual place the tube in the safe-non-destruct regions of the tube characteristics, as long as the operating potentials are observed.

4.7-4E   RESISTANCES

Two resistive values listed in tube manuals are plate resistance (R_p) and sometimes suggested load resistance (R_L).

4.7-4E1.   Plate Resistance—The R_p of vacuum tubes is actually an application of Ohms Law since it is a ratio of E (voltage) divided by I (current). The E is the small change in plate voltage, while I is the corresponding change in plate current when all other electrodes are held constant.

4.7-4E2   Load Resistance—Power amplifier stage vacuum tubes normally have suggested load resistances listed to insure maximum wattage utilization at average or maximum value operation.

4.7-4F   VOLTAGES

The voltages listed in tube manual data are pretty much "matter of fact" values since tube characteristics were taken into consideration. The data strongly suggested for plate and screen voltage are either average or maximum values and should not be exceeded. This, therefore, means other values for plate voltages may be used as long as maximum plate voltage ratings are not exceeded, and other screen grid potentials may be used providing the maximum voltage is not exceeded. Limitations for screen grid current should, however, be observed when maximum plate voltage is employed.

The grid bias, or negative grid voltage potentials, are so chosen in order to provide plate and screen currents while holding plate and screen dissipations wtihin maximum values, thereby providing satisfactory tube operation.

Filament voltages may be operated with AC or DC voltages unless specifically noted.

CHAPTER 5

Time-Saving Tables, Measurements, and Service Data

5.1   TRANSISTOR ALPHA-BETA RELATIONSHIPS

The common PNP and NPN alpha-beta gain relationships may seem confusing to those of us who have not yet realized that this relationship portrays current gains for two different circuitry configurations.

The first circuit configuration is called a grounded or common base and is characterized by a low input impedance and a high output impedance. The current gain for this type of circuit is called alpha (α) and is always less than one.

The second circuit configuration, called a grounded or common emitter, provides what is called an intermediate input impedance (about 1 kilohm) and an intermediate output impedance (about 50 kilohm). The current gain for this particular circuit and also the common collector circuit is called beta (β) and will always be greater than unity; in fact gains higher than 200 are possible and normal.

Most transistor literature hints as to circuitry configuration when telling the current gain for a specific transistor. The current gain value is normally listed under the letters HFE or under the heading Direct Current Gain. These values simply relate a number obtained by ratios of collector and base currents for a common emitter configuration or by ratios of collector and emitter currents for a common base configuration. In essence it tells the ability of the transistor to amplify currents flowing through its elements.

Since alpha and beta have formulas indicating their relationship:

$$\alpha = \frac{\beta}{1 + \beta} \quad \text{AND} \quad \beta = \frac{\alpha}{1 - \alpha}$$

you could calculate one or the other, if one were known. If you found a number less than one under a transistor's current gain (alpha) or a number greater than one (beta), the other value could be obtained using the formulas, if circuitry demanded. Most of these common alpha-beta relationship values are shown here to lessen your calculation time.

BETA	ALPHA	BETA	ALPHA
1	0.5000	4	0.8000
2	0.6666	5	0.8333
3	0.7500	6	0.8571

BETA	ALPHA	BETA	ALPHA
7	0.8750	53	0.9815
8	0.8889	54	0.9818
9	0.9000	55	0.9821
10	0.9091	56	0.9825
11	0.9167	57	0.9828
12	0.9231	58	0.9831
13	0.9286	59	0.9833
14	0.9333	60	0.9836
15	0.9375	61	0.9839
16	0.9412	62	0.9841
17	0.9444	63	0.9844
18	0.9474	64	0.9846
19	0.9500	65	0.9848
20	0.9524	66	0.9851
21	0.9545	67	0.9853
22	0.9565	68	0.9855
23	0.9583	69	0.9857
24	0.9600	70	0.9859
25	0.9615	71	0.9861
26	0.9630	72	0.9863
27	0.9643	73	0.9865
28	0.9655	74	0.9867
29	0.9667	75	0.9868
30	0.9677	76	0.9870
31	0.9688	77	0.9872
32	0.9697	78	0.9873
33	0.9706	79	0.9875
34	0.9714	80	0.9877
35	0.9722	81	0.9878
36	0.9730	82	0.9880
37	0.9737	83	0.9881
38	0.9744	84	0.9882
39	0.9750	85	0.9884
40	0.9756	86	0.9885
41	0.9762	87	0.9886
42	0.9767	88	0.9888
43	0.9773	89	0.9889
44	0.9778	90	0.9890
45	0.9783	91	0.9891
46	0.9787	92	0.9892
47	0.9792	93	0.9894
48	0.9796	94	0.9895
49	0.9800	95	0.9896
50	0.9804	96	0.9897
51	0.9808	97	0.9898
52	0.9811	98	0.9899

BETA	ALPHA	BETA	ALPHA
99	0.9900	280	0.9964
100	0.9901	290	0.9966
110	0.9910	300	0.99667
120	0.9917		
130	0.9924		
140	0.9929		
150	0.9934		
160	0.9938		
170	0.9942		
180	0.9945		
190	0.9948		
200	0.9950		
210	0.9953		
220	0.9955		
230	0.9957		
240	0.9959		
250	0.9960		
260	0.9962		
270	0.9963		

5.2   AMERICAN WIRE GAGE BARE WIRE SPECIFICATIONS

Wire Size (AWG) (B+S)	Nominal Diameter in Mils $\left(\frac{1}{1000} \text{ of inch}\right)$	*Diameter in MM.	Circular Mil-Area	OHMS Per 1000 Ft. @ 68° F (20° C)	Current Carrying Capacity @ 700 CM/AMP
0000	460.0	11.684	211,600	0.0490	302.3
000	409.6	10.404	167,800	0.0618	239.7
00	364.8	9.266	133,100	0.0779	190.1
0	324.9	8.252	105,500	0.0983	150.9
1	289.3	7.348	83,690	0.1239	119.6
2	257.6	6.543	66,370	0.1563	94.8
3	229.4	5.827	52,640	0.1970	75.2
4	204.3	5.189	41,740	0.2485	59.6
5	181.9	4.620	33,100	0.3133	47.3
6	162.0	4.115	26,250	0.3951	37.5
7	144.3	3.665	20,820	0.4982	29.7
8	128.5	3.264	16,510	0.6282	23.6
9	114.4	2.906	13,090	0.7921	18.7
10	101.9	2.588	10,380	0.9989	14.8
11	90.74	2.305	8,234	1.260	11.8
12	80.81	2.053	6,530	1.588	9.33

*Rounded off to third significant digit using 0.0254 as the conversion factor.

Wire Size (AWG) (B+S)	Nominal Diameter in Mils $\left(\dfrac{1 \text{ of inch}}{1000}\right)$	*Diameter in MM.	Circular Mil-Area	OHMS Per 1000 Ft. @ 68° F (20° C)	Current Carrying Capacity @ 700 CM/AMP
13	71.96	1.828	5,178	2.003	7.40
14	64.08	1.628	4,107	2.525	5.87
15	57.07	1.450	3,257	3.184	4.65
16	50.82	1.291	2,583	4.016	3.69
17	45.26	1.150	2,048	5.064	2.93
18	40.30	1.024	1,624	6.385	2.32
19	35.89	0.912	1,288	8.051	1.84
20	31.96	0.812	1,022	10.15	1.46
21	28.46	0.723	810.1	12.80	1.16
22	25.35	0.644	642.4	16.14	0.918
23	22.57	0.573	509.5	20.36	0.728
24	20.10	0.511	404.0	25.67	0.577
25	17.90	0.455	320.4	32.37	0.458
26	15.94	0.405	254.1	40.81	0.363
27	14.20	0.361	201.5	51.47	0.288
28	12.64	0.321	159.8	64.90	0.228
29	11.26	0.286	126.7	81.83	0.181
30	10.03	0.255	100.5	103.2	0.144
31	8.928	0.227	79.70	130.1	0.114
32	7.950	0.202	63.21	164.1	0.090
33	7.080	0.180	50.13	206.9	0.072
34	6.305	0.160	39.75	260.9	0.057
35	5.615	0.143	31.52	329.0	0.045
36	5.000	0.127	25.00	414.8	0.036
37	4.453	0.113	19.83	523.1	0.028
38	3.965	0.101	15.72	659.6	0.022
39	3.531	0.090	12.47	831.8	0.018
40	3.145	0.080	9.89	1049.0	0.014

5.3   DECIBELS

GAIN

Decibel	E/I Ratios	Power Ratios
0	1.000	1.000
0.2	1.023	1.047
0.4	1.047	1.096
0.6	1.072	1.148
0.8	1.096	1.202
1.0	1.122	1.259

LOSS

Decibel	E/I Ratios	Power Ratios
0	1.0000	1.0000
−0.2	0.9772	0.9550
−0.4	0.9550	0.9120
−0.6	0.9333	0.8710
−0.8	0.9120	0.8318
−1.0	0.8913	0.7943

GAIN

Decibel	E/I Ratios	Power Ratios
1.2	1.148	1.318
1.4	1.175	1.380
1.6	1.202	1.445
1.8	1.230	1.514
2.0	1.259	1.585
2.2	1.288	1.660
2.4	1.318	1.738
2.6	1.349	1.820
2.8	1.380	1.905
3.0	1.413	1.995
3.2	1.445	2.089
3.4	1.479	2.188
3.6	1.514	2.291
3.8	1.549	2.399
4.0	1.585	2.512
4.2	1.622	2.630
4.4	1.660	2.754
4.6	1.698	2.884
4.8	1.738	3.020
5.0	1.778	3.162
5.2	1.820	3.311
5.4	1.862	3.467
5.6	1.905	3.631
5.8	1.950	3.802
6.0	1.995	3.981
6.2	2.042	4.169
6.4	2.089	4.365
6.6	2.138	4.571
6.8	2.188	4.786
7.0	2.239	5.012
7.2	2.291	5.248
7.4	2.344	5.495
7.6	2.399	5.754
7.8	2.455	6.026
8.0	2.512	6.310
8.2	2.570	6.607
8.4	2.630	6.918
8.6	2.692	7.244
8.8	2.754	7.586
9.0	2.818	7.943
9.2	2.884	8.318
9.4	2.951	8.710
9.6	3.020	9.120

LOSS

Decibel	E/I Ratios	Power Ratios
−1.2	0.8710	0.7586
−1.4	0.8511	0.7244
−1.6	0.8318	0.6918
−1.8	0.8128	0.6607
−2.0	0.7943	0.6310
−2.2	0.7762	0.6026
−2.4	0.7586	0.5754
−2.6	0.7413	0.5495
−2.8	0.7244	0.5248
−3.0	0.7079	0.5012
−3.2	0.6918	0.4786
−3.4	0.6761	0.4571
−3.6	0.6607	0.4365
−3.8	0.6457	0.4169
−4.0	0.6310	0.3981
−4.2	0.6166	0.3802
−4.4	0.6026	0.3631
−4.6	0.5888	0.3467
−4.8	0.5754	0.3311
−5.0	0.5623	0.3162
−5.2	0.5495	0.3020
−5.4	0.5370	0.2884
−5.6	0.5248	0.2754
−5.8	0.5129	0.2630
−6.0	0.5012	0.2512
−6.2	0.4898	0.2399
−6.4	0.4786	0.2291
−6.6	0.4677	0.2188
−6.8	0.4571	0.2089
−7.0	0.4467	0.1995
−7.2	0.4365	0.1905
−7.4	0.4266	0.1820
−7.6	0.4169	0.1738
−7.8	0.4074	0.1660
−8.0	0.3981	0.1585
−8.2	0.3890	0.1514
−8.4	0.3802	0.1445
−8.6	0.3715	0.1380
−8.8	0.3631	0.1318
−9.0	0.3548	0.1259
−9.2	0.3467	0.1202
−9.4	0.3388	0.1148
−9.6	0.3311	0.1096

GAIN

LOSS

Decibel	E/I Ratios	Power Ratios	Decibel	E/I Ratios	Power Ratios
9.8	3.090	9.550	−9.8	0.3236	0.1047
10.0	3.162	10.00	−10.0	0.3162	0.1000
10.2	3.236	10.47	−10.2	0.3090	0.0955
10.4	3.311	10.96	−10.4	0.3020	0.0912
10.6	3.388	11.48	−10.6	0.2951	0.0871
10.8	3.467	12.02	−10.8	0.2884	0.0832
11.0	3.548	12.59	−11.0	0.2818	0.0794
11.2	3.631	13.18	−11.2	0.2754	0.0759
11.4	3.715	13.80	−11.4	0.2692	0.0724
11.6	3.802	14.45	−11.6	0.2630	0.0692
11.8	3.890	15.14	−11.8	0.2570	0.0661
12.0	3.981	15.85	−12.0	0.2512	0.0631
12.2	4.074	16.60	−12.2	0.2455	0.0603
12.4	4.169	17.38	−12.4	0.2399	0.0575
12.6	4.266	18.20	−12.6	0.2344	0.0550
12.8	4.365	19.05	−12.8	0.2291	0.0525
13.0	4.467	19.95	−13.0	0.2239	0.0501
13.2	4.571	20.89	−13.2	0.2188	0.0479
13.4	4.677	21.88	−13.4	0.2138	0.0457
13.6	4.786	22.91	−13.6	0.2089	0.0437
13.8	4.898	23.99	−13.8	0.2042	0.0417
14.0	5.012	25.12	−14.0	0.1995	0.0398
14.2	5.129	26.30	−14.2	0.1950	0.0380
14.4	5.248	27.54	−14.4	0.1905	0.0363
14.6	5.370	28.84	−14.6	0.1862	0.0347
14.8	5.495	30.20	−14.8	0.1820	0.0331
15.0	5.623	31.62	−15.0	0.1778	0.0316
15.2	5.754	33.11	−15.2	0.1738	0.0302
15.4	5.888	34.67	−15.4	0.1698	0.0288
15.6	6.026	36.31	−15.6	0.1660	0.0275
15.8	6.166	38.02	−15.8	0.1622	0.0263
16.0	6.310	39.81	−16.0	0.1585	0.0251
16.2	6.457	41.69	−16.2	0.1549	0.0240
16.4	6.607	43.65	−16.4	0.1514	0.0229
16.6	6.761	45.71	−16.6	0.1479	0.0219
16.8	6.918	47.86	−16.8	0.1445	0.0209
17.0	7.079	50.12	−17.0	0.1413	0.0199
17.2	7.244	52.48	−17.2	0.1380	0.0191
17.4	7.413	54.95	−17.4	0.1349	0.0182
17.6	7.586	57.54	−17.6	0.1318	0.0174
17.8	7.762	60.26	−17.8	0.1288	0.0166
18.0	7.943	63.10	−18.0	0.1259	0.0159
18.2	8.128	66.07	−18.2	0.1230	0.0151

GAIN				LOSS		
Decibel	E/I Ratios	Power Ratios		Decibel	E/I Ratios	Power Ratios
18.4	8.318	69.18		−18.4	0.1202	0.0145
18.6	8.511	72.44		−18.6	0.1175	0.0138
18.8	8.710	75.86		−18.8	0.1148	0.0132
19.0	8.913	79.43		−19.0	0.1122	0.0126
19.2	9.120	83.18		−19.2	0.1096	0.0120
19.4	9.333	87.10		−19.4	0.1072	0.0115
19.6	9.550	91.20		−19.6	0.1047	0.0110
19.8	9.772	95.50		−19.8	0.1023	0.0105
20.0	10.00	100.00		−20.0	0.1000	0.0100
25.0	17.78	316.2		−25.0	0.0562	0.00316
30.0	31.62	1000.0		−30.0	0.0316	0.00100
35.0	56.23	3162.0		−35.0	0.0178	3.162×10^{-4}
40.0	100.00	10,000.0		−40.0	0.0100	1×10^{-4}
45.0	177.8	31,620.0		−45.0	0.0056	3.162×10^{-5}
50.0	316.2	100,000.0		−50.0	0.0032	1×10^{-5}

5.4   DRILL, TAP, AND MACHINE SCREW SPECIFICS

PLUS DECIMAL EQUIVALENTS

Tap Drill	Drill Diameter	Machine Screw	Tap Drill	Drill Diameter	Machine Screw
97	0.0059	———	77	0.0180	———
96	0.0063	———	76	0.0200	———
95	0.0067	———	75	0.0210	———
94	0.0071	———	74	0.0225	———
93	0.0075	———	73	0.0240	———
92	0.0079	———	72	0.0250	———
91	0.0083	———	71	0.0260	———
90	0.0087	———	70	0.0280	———
89	0.0091	———	69	0.0292	———
88	0.0095	———	68	0.0310	———
87	0.0100	———	1/32	0.0313	———
86	0.0105	———	67	0.0320	———
85	0.0110	———	66	0.0330	———
84	0.0115	———	65	0.0350	———
83	0.0120	———	64	0.0360	———
82	0.0125	———	63	0.0370	———
81	0.0130	———	62	0.0380	———
80	0.0135	———	61	0.0390	———
79	0.0145	———	60	0.0400	———
1/64	0.0156	———	59	0.0410	———
78	0.0160	———	58	0.0420	———

5.4　DRILL, TAP, AND MACHINE SCREW SPECIFICS
PLUS DECIMAL EQUIVALENTS *(Continued)*

Tap Drill	Drill Diameter	Machine Screw	Tap Drill	Drill Diameter	Machine Screw
57	0.0430	———	28	0.1405	8-40
56	0.0465	———	9/64	0.1406	11/64-32
3/64	0.0469	0-80	27	0.1440	9-30
		1/16-64	26	0.1470	9-32, 3/16-24
		1/16-72	25	0.1495	10-24
55	0.0520	———	24	0.1520	———
54	0.0550	1-56	23	0.1540	10-28
53	0.0595	1-64	5/32	0.1562	———
		1-72	22	0.1570	10-30, 3/16-32
1/16	0.0625	5/64-60	21	0.1590	10-32
52	0.0635	5/64-72	20	0.1610	13/64-24
51	0.0670	———	19	0.1660	
50	0.0700	2-56, 2-64	18	0.1695	———
49	0.0730	3/32-48	11/64	0.1719	———
		3/32-50	17	0.1730	———
48	0.0760	———	16	0.1770	12-24,
5/64	0.0781	———			7/32-24
47	0.0785	3-48	15	0.1800	———
46	0.0810	———	14	0.1820	12-28
45	0.0820	3-56, 4-32	13	0.1850	12-32
44	0.0860	4-36	3/16	0.1875	———
43	0.0890	4-40, 7/64-48	12	0.1890	7/32-32
42	0.0935	4-48	11	0.1910	
3/32	0.0938	1/8-32	10	0.1935	14-20
41	0.0960	———			15/64-24
40	0.0980	5-36	9	0.1960	———
39	0.0995	———	8	0.1990	———
38	0.1015	5-40, 1/8-40	7	0.2010	14-24, ¼-20
37	0.1040	5-44	13/64	0.2031	———
36	0.1065	6-32	6	0.2040	———
7/64	0.1094	———	5	0.2055	———
35	0.1100	———	4	0.2090	¼-24
34	0.1110	6-36	3	0.2130	16-18, ¼-27
					¼-28
33	0.1130	6-40	7/32	0.2188	16-20, ¼-32
32	0.1160	9/64-40	2	0.2210	16-22
31	0.1200	7-30, 7-32	1	0.2280	———
1/8	0.1250	7-36, 5/32-32	A	0.2340	———
30	0.1285	8-30, 5/32-36			
			15/64	0.2344	———
29	0.1360	8-32, 8-36, 9-24	B	0.2380	18-18

5.4   DRILL, TAP, AND MACHINE SCREW SPECIFICS

PLUS DECIMAL EQUIVALENTS *(Continued)*

Tap Drill	Drill Diameter	Machine Screw	Tap Drill	Drill Diameter	Machine Screw
C	0.2420	———	7/16	0.4375	———
D	0.2460	18-20	29/64	0.4531	½-20
					½-24
E, ¼	0.2500	———	15/32	0.4688	½-27
F	0.2570	5/16-18	31/64	0.4844	9/16-12
G	0.2610	20-16	1/2	0.5000	———
17/64	0.2656	20-18, 5/16-20	33/64	0.5156	9/16-18
H	0.2660	———	17/32	0.5312	9/16-27
I	0.2720	20-20, 5/16-24			5/8-11
J	0.2770	5/16-27	35/64	0.5469	5/8-12
			9/16	0.5625	———
K	0.2810		37/64	0.5781	5/8-18
9/32	0.2812	22-16, 5/16-32	19/32	0.5938	5/8-27
L	0.2900	22-18			11/16-11
M	0.2950	———	39/64	0.6094	———
19/64	0.2969	———	5/8	0.6250	11/16-16
N	0.3020	———	41/64	0.6406	———
5/16	0.3125	24-16, 3/8-16	21/32	0.6562	3/4-10
0	0.3160	24-18	43/64	0.6719	3/4-12
			11/16	0.6875	3/4-16
P	0.3230		45/64	0.7031	———
21/64	0.3281	26-14, 3/8-20	23/32	0.7188	3/4-27
Q	0.3320	3/8-24			13/16-10
R	0.3390	26-16	47/64	0.7344	———
		3/8-27	3/4	0.7500	———
11/32	0.3438	———	49/64	0.7656	7/8-9
S	0.3480	———	25/32	0.7812	———
T	0.3580	28-14	51/64	0.7969	7/8-12
23/64	0.3594	28-16			
U	0.3680	7/16-14	13/16	0.8125	7/8-14
3/8	0.3750	———	53/64	0.8281	7/8-18
V	0.3770	30-14			15/16-9
W	0.3860	———	27/32	0.8438	7/8-27
25/64	0.3906	30-16,	55/64	0.8594	———
		7/16-20	7/8	0.8750	1-8
X	0.3970	7/16-24	57/64	0.8906	———
Y	0.4040	7/16-27	29/32	0.9062	———
13/32	0.4062	———			
Z	0.4130	———	59/64	0.9219	1-12
27/64	0.4219	½-12	15/16	0.9375	1-14
		½-13	61/64	0.9531	———

5.4 DRILL, TAP, AND MACHINE SCREW SPECIFICS
PLUS DECIMAL EQUIVALENTS *(Continued)*

Tap Drill	Drill Diameter	Machine Screw
31/32	0.9688	1-27
63/64	0.9844	———
1	1.000	———

5.5 DECIMAL, MILLIMETER, FRACTIONAL UNIT OF AN INCH EQUIVALENTS

FRACTION			DECIMAL	*MILLIMETER
		1/64	0.015625	0.3969
	1/32		0.03125	0.7938
		3/64	0.046875	1.1906
1/16			0.0625	1.5875
		5/64	0.078125	1.9844
	3/32		0.09375	2.3813
		7/64	0.109375	2.7781
1/8			0.125	3.1750
		9/64	0.140625	3.5719
	5/32		0.15625	3.9688
		11/64	0.171875	4.3656
3/16			0.1875	4.7625
		13/64	0.203125	5.1594
	7/32		0.21875	5.5563
		15/64	0.234375	5.9531
1/4			0.250	6.3500
		17/64	0.265625	6.7469
	9/32		0.28125	7.1438
		19/64	0.296875	7.5406
5/16			0.3125	7.9375
		21/64	0.328125	8.3344
	11/32		0.34375	8.7313
		23/64	0.359375	9.1281
3/8			0.375	9.5250
		25/64	0.390625	9.9219
	13/32		0.40625	10.3188
		27/64	0.421875	10.7156
7/16			0.4375	11.1125

*Using internationally agreed conversion factor of 25.4 and rounding off last digit if equaling 5 or greater.

5.5　DECIMAL, MILLIMETER, FRACTIONAL UNIT OF AN INCH EQUIVALENTS
(Continued)

FRACTION			DECIMAL	*MILLIMETER
		29/64	0.453125	11.5094
	15/32		0.46875	11.9063
		31/64	0.484375	12.3031
1/2			0.500	12.7000
		33/64	0.515625	13.0969
	17/32		0.53125	13.4938
		35/64	0.546875	13.8906
	9/16		0.5625	14.2875
		37/64	0.578125	14.6844
	19/32		0.59375	15.0813
		39/64	0.609375	15.4781
5/8			0.625	15.8750
		41/64	0.640625	16.2719
	21/32		0.65625	16.6688
		43/64	0.671875	17.0656
	11/16		0.6875	17.4625
		45/64	0.703125	17.8594
	23/32		0.71875	18.2563
		47/64	0.734375	18.6531
3/4			0.750	19.0500
		49/64	0.76525	19.4469
	25/32		0.78125	19.8438
		51/64	0.796875	20.2406
	13/16		0.8125	20.6375
		53/64	0.828125	21.0344
	27/32		0.843750	21.4313
		55/64	0.859375	21.8281
7/8			0.875	22.2250
		57/64	0.890625	22.6219
	29/32		0.90625	23.0188
		59/64	0.921875	23.4156
	15/16		0.9375	23.8125
		61/64	0.953125	24.0938
	31/32		0.96875	24.6063
		63/64	0.984375	25.0031
1			1.000	25.4000

5.6　dBmV CONVERSION TABLE
(Reverence Level: 0 dBmV = 1000 μV = 1mV)

dBmV	micro-volt (μV)	dBmV	millivolt (mV)	dBmV	millivolt (mV)
−40	10.00	0	1.000	41	112.20
−39	11.22	1	1.122	42	125.90

5.6   dBmV CONVERSION TABLE
(Continued)

dBmV	Micro-volt (μV)	dBmV	Millivolt (mV)	dBmV	millivolt (mV)
−38	12.59	2	1.259	43	141.30
−37	14.13	3	1.413	44	158.50
−36	15.85	4	1.585	45	177.80
−35	17.78	5	1.778	46	199.50
−34	19.95	6	1.995	47	223.90
−33	22.39	7	2.239	48	251.20
−32	25.12	8	2.512	49	281.80
−31	28.18	9	2.818	50	316.20
−30	31.62	10	3.162	51	354.80
−29	35.48	11	3.548	52	398.10
−28	39.81	12	3.981	53	446.70
−27	44.67	13	4.467	54	501.20
−26	50.12	14	5.012	55	562.30
−25	56.23	15	5.623	56	631.00
−24	63.10	16	6.310	57	707.90
−23	70.79	17	7.079	58	794.30
−22	79.43	18	7.943	59	891.30
−21	89.13	19	8.913	60	1,000.00
−20	100.0	20	10.000	61	1,1220.00
−19	112.2	21	11.220	62	1,259.00
−18	125.9	22	12.590	63	1,413.00
−17	141.3	23	14.130	64	1,585,00
−16	158.5	24	15.850	65	1,778.00
−15	177.8	25	17.780	66	1,995.00
−14	199.5	26	19.950	67	2,239.00
−13	223.9	27	22.390	68	2,512.00
−12	251.2	28	25.120	69	2,818.00
−11	281.8	29	28.180	70	3,162.00
−10	316.2	30	31.620	71	3,548.00
−9	354.8	31	35.480	72	3,981.00
−8	398.1	32	39.810	73	4,467.00
−7	446.7	33	44.670	74	5,012.00
−6	501.2	34	50.120	75	5,632.00
−5	562.3	35	56.230	76	6,310.00
−4	631.0	36	63.100	77	7,079.00
−3	707.9	37	70.790	78	7,943.00
−2	794.3	38	79.430	79	8,913.00
−1	891.3	39	89.130	80	10,000.00
0	1,000.0	40	100.00		

5.7   GAS-FILLED LAMPS

The following neon, argon, and circuit glow lamps arranged according to their normal working voltage range show the current ratings, series resistance, if any required, and their popular lamp number. Some of them have two identification numbers illustrated. The second group shows the American Standards Association (ASA) number. When ordering, any listed number may be used.

Some of the similar lamp base style terminology common to gas-filled and miniature lamps is provided in Figure 5-1 for your reference.

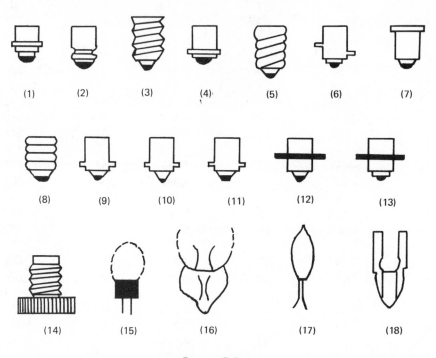

Figure 5-1

1. Submidget flanged	7. Miniature flange	13. Double contact prefocus
2. Midget grooved	8. Miniature screw	14. Knurled screw
3. Midget screw	9. Miniature bayonet	15. Bi-pin
4. Midget flanged	10. Single contact bayonet	16. Wedge
5. Candelabra	11. Double contact bayonet	17. Wire terminal
6. Double contact bayonet index	12. Single contact prefocus	18. Telephone slide

Voltage Range	Lamp Number	Current Rating (mA)	Series Resistance	Base
55-90 volts	NE-3 (8AA)	0.3 mA	—	Tel. slide
	NE-67 (6AC)	0.2 mA	—	min bay
	NE-86 (5AJ)	1.5 mA	—	wire term
	5AJA	1.5 mA	—	wire term
60-65 volts	NE-83 (5AH)	10 mA	—	wire term

Voltage Range	Lamp Number	Current Rating (mA)	Series Resistance	Base
60-80 volts	3AGA	0.4 mA	—	wire term
	5AB-A	0.3 mA	—	wire term
60-85 volts	5AH-A	10 mA	—	wire term
60-90 volts	NE-4 (8AB)	0.3 mA	—	tel. slide
	NE-23 (5AB)	0.3 mA	—	wire term
	NE-68 (5AC)	0.3 mA	—	wire term
	3AG	0.4 mA	—	wire term
60-100 volts	8AE	10 mA	—	tel. slide
62-72 volts	5AB-B	0.3 mA	—	wire term.
64-80 volts	NE-81 (5AG)	0.1 mA	—	wire term.
65-73 volts	3AG-B	0.4 mA	—	wire term.
67-87 volts	NE-16 (7AA)	1.5 mA	—	D.C. bay.
68-76 volts	NE-76 (5AG-A)	0.4 mA	—	wire term.
	3AG-C	0.4 mA	—	wire term.
68-78 volts	5AB-C	0.3 mA	—	wire term.
70-90 volts	5AJ-B	1.5 mA	—	wire term.
75-85 volts	5AH-B	10 mA	—	wire term.
75-100 volts	5AH-C	10 mA	—	wire term.
75-135 volts	3AH	2 mA	—	wire term.
105-125 volts	NE-2 (A1A)	0.5 mA	150k ohms	wire term.
	A1B	0.3 mA	220k ohms	wire term.
	A1C	0.8 mA	68k ohms	wire term.
	NE-2A (A2A)	0.3 mA	220k ohms	wire term.
	NE-2D (C7A)	0.7 mA	100k ohms	s.c. midget-flg.
	NE-2E (A9A)	0.7 mA	100k ohms	wire term.
	NE-2H (C2A)	1.9 mA	30k ohms	wire term.
	NE-2J (C9A)	1.9 mA	30k ohms	S.C. midget-flg.
	NE-2V (A2B)	0.7 mA	100k ohms	wire term.
	NE-7 (B4A)	2.0 mA	30k ohms	wire term.
	NE-17 (B5A)	2.0 mA	30k ohms	D.C. bay.
	NE-21 (B6A)	2.0 mA	30k ohms	S.C. bay.
	NE-30 (J5A)	12.0 mA	4.8k ohms	med. scr.
	NE-32(L6A)(L5A)	12.0 mA	7.5k ohms	D.C. bay.
	NE-45 (B7A)	2.0 mA	30k ohms	cand. scr.
	NE-47 (B8A)	2.0 mA	30K ohms	S.C. bay.
	NE-48 (B9A)	2.0 mA	30k ohms	D.C. bay.
	NE-51 (B1A)	0.3 mA	220k ohms	min. bay.
	NE-51H (B2A)	1.2 mA	47k ohms	min. bay.
	NE-54 (F2A)	2.0 mA	30k ohms	wire term.
	NE-57 (F3A)	2.0 mA	30k ohms	cand. scr.
	NE-66	1.0 mA	3.6k ohms	cand. scr.
	NE-79	12.0 mA	7.5k ohms	bay.
	(J2A)(AR3)	3.5 mA	15k ohms	cand. scr.
	(J3A)(AR4)	3.5 mA	15k ohms	D.C. bay.

Voltage Range	Lamp Number	Current Rating	Series Resistance	Base
110-125 volts	AR1	18 mA	—	med. scr.
	AR3	3.5 mA	—	cand. scr.
	AR4	3.5 mA	15k ohms	D.C. bay.
	NE-34	18 mA	—	med. scr.
	NE-40	30 mA	—	med. scr.
110-140 volts	NE-97 (4AC)	0.5 mA	—	wire term.
120-150 volts	NE-96 (4AB)	0.5 mA	—	wire term.
	NE-5 (8AC)	0.5 mA	—	tel. slide
210-250 volts	NE-56 (J9A)	5 mA	30k ohms	med. scr.
	NE-58 (F4A)	2 mA	100k ohms	cand. scr.
220-300 volts	AR1	18 mA	10k ohms	med. scr.
	AR3	3.5 mA	68k ohms	cand. scr.
	AR4	3.5 mA	82k ohms	D.C. bay.
	NE-2	3 mA	750k ohms	wire term.
	NE-2A	3 mA	750k ohms	wire term.
	NE-17	2 mA	110k ohms	D.C. bay.
	NE-30	12 mA	10k ohms	med. scr.
	NE-32	12 mA	18k ohms	D.C. bay.
	NE-34	18 mA	9.1k ohms	med. scr.
	NE-40	30 mA	6.2k ohms	med. scr.
	NE-45	2 mA	82k ohms	cand. scr.
	NE-48	2 mA	110k ohms	D.C. bay.
	NE-51	0.3 mA	750k ohms	min. bay.
	NE-57	2 mA	82k ohms	cand. scr.
300-375 volts	AR1	18 mA	18k ohms	med. scr.
	AR3	3.5 mA	9.1k ohms	cand. scr.
	AR4	3.5 mA	100k ohms	D.C. bay.
	NE-2	3 mA	1M ohms	wire term.
	NE-2A	3 mA	1M ohms	wire term.
	NE-17	2 mA	150k ohms	D.C. bay.
	NE-30	12 mA	20k ohms	med. scr.
	NE-32	12 mA	27k ohms	D.C. bay.
	NE-34	18 mA	13k ohms	med. scr.
	NE-40	30 mA	8.2k ohms	med. scr.
	NE-45	2 mA	120k ohms	cand. scr.
	NE-48	2 mA	150k ohms	D.C. bay.
	NE-51	0.3 mA	1M ohms	min. bay.
	NE-57	2 mA	120k ohms	cand. scr.
375-450 volts	AR1	18 mA	24k ohms	med. scr.
	AR3	3.5 mA	150k ohms	cand. scr.
	AR4	3.5 mA	160k ohms	D.C. bay.
	NE-2	3 mA	1.2M ohms	wire term.
	NE-2A	3 mA	1.2M ohms	wire term.
	NE-17	2 mA	180k ohms	D.C. bay.

Voltage Range	Lamp Number	Current Rating	Series Resistance	Base
375-450 volts	NE-30	12 mA	24k ohms	med. scr.
	NE-32	12mA	33k ohms	D.C. bay.
	NE-34	18 mA	16k ohms	med. scr.
	NE-40	30 mA	11k ohms	med. scr.
	NE-45	2mA	150k ohms	cand. scr.
	NE-48	2 mA	180k ohms	D.C. bay.
	NE-51	0.3 mA	1.2M ohms	min. bay.
	NE-57	2 mA	150k ohms	cand. scr.
450-600 volts	AR1	18 mA	30k ohms	Med. scr.
	AR3	3.5 mA	160k ohms	Cand. scr.
	AR4	3.5 mA	180k ohms	D.C. bay.
	NE-2	3 mA	1.6M ohms	wire term.
	NE-2A	3 mA	1.6M ohms	wire term.
	NE-17	2 mA	240k ohms	D.C. bay.
	NE-30	12 mA	36k ohms	med. scr.
	NE-32	12 mA	43k ohms	D.C. bay.
	NE-34	18 mA	22k ohms	med. scr.
	NE-40	30 mA	16k ohms	med. scr.
	NE-45	2 mA	200k ohms	cand. scr.
	NE-48	2 mA	240k ohms	D.C. bay.
	NE-51	0.3 mA	1.6M ohms	min. bay.
	NE-57	2 mA	200k ohms	cand. scr.

5.8 MINIATURE LAMP DATA

Throughout the field of electronics one will find employment of the common miniature lamps used so often as indicating devices. The word miniature might be misleading; it does not relate to a specific range of sizes, but does usually mean its operating voltage is under 60 volts.

Miniature lamp size, shape, base type, and number-letter combination have, for the most part, become standardized. The following letter and number bulb identification tells us from the letter the shape of the bulb, while the number indicates the approximate bulb diameter in eighths of an inch. For example:

Bulb Shape

(B)

Figure 5-2

Number-Letter Combination	Approximate Diameter
B-3½	7/16
B-6	3/4
G-2	1/4
G-3½	7/16

Bulb Shape	Number-Letter Combination	Approximate Diameter
	G-4½	9/16
Figure 5-3	G-5	5/8
	G-5½	3/4
	G-6	3/4
(G)	G-8	1
	G-16½	2 1/16
Figure 5-4	R-12	1½
(R)	RP-11	1 3/8
	S-8	1
Figure 5-5	S-11	1 3/8
	T-3/4	3/32
(RP)	T-1	1/8
	T-1¼	5/32
	T-1 3/4	7/32
Figure 5-6	T-2	1/4
(S)	T-3	3/8
Figure 5-7	T-3¼	13/32
	T-4	1/2
(T)	T-4½	9/16
	T-5	5/8

Miniature lamp manufacturers make several different styles and sizes of base lamps. Twenty-two different base types are listed throughout the following organized miniature lamp data. The base types referred to are:

Base Type	Abbreviation
Bi pin	————
Candelabra screw	Cand scr
Double contact baynet	D.C. bay
Double contact index	D.C. index
Double contact prefocus	D.C. prefocus
Knurled screw	Knurled scr.
Micro-midget flange	Micro-midget flg.
Midget flange	Midget flg.
Midget groove	————
Miniature baynet	Min bay
Miniature screw	Min scr
Miniature 2 pin	Min 2 pin (bi-pin)
Short slide	————
Single contact baynet	S.C. bay
Single contact miniature flange	S.C. min. flg.
Single contact prefocus	S.C. prefocus
Special midget flange	————
Special midget screw	Special midget scr

Base Type	Abbreviation
Special screw	Special scr
Sub-midget flange	Sub-midget flg
Telephone slide	————
Wire terminal	Wire term

In the following lamp data, each lamp number has been grouped under its specific operating voltage range. The current, maximum overall length (mol), bulb type, and base type or style have also been listed to aid you in finding or substituting miniature lamps to suit your immediate needs. Each bulb number will have its manufacturer's specific letter combination assigned to it when you buy their brand. For example, if General Electric were bought, then GE-47, Chicago miniature CM-47, etc. However, technical data remains as shown.

VOLTAGE	LAMP #	CURRENT	MOL	BULB TYPE	BASE
1.15v	8712	0.90A	0.187	T-¾	wire term
1.25v	123	0.30A	0.940	G-3½	min scr
	136	0.60A	1.060	G-4½	min scr
	423	0.60A	1.060	G-4½	min bay
	7244	0.012A	0.375	T-1	sub midget flg
	7252	0.012A	0.380	T-1	Bi pin
	8798	0.012A	0.250	T-1	wire term
1.3v	131	0.100A	0.940	G-3½	min scr
	2135	0.030A	0.460	T-1¼	wire term
	7301	0.030A	0.546	T-1¼	knurled scr
	7302	0.030A	0.500	T-1¼	special midget scr
	7303	0.030A	0.550	T-1¼	special scr
	7306	0.030A	0.600	T-1¼	Bi pin
1.35v	331	0.060A	0.625	T-1¾	midget flg
	359	0.060A	0.940	G-3½	min scr
	698	0.060A	0.625	T-1¾	midget groove
	1728	0.060A	0.520	T-1¾	wire term
	1800	0.060A	1.190	T-3¼	min scr
	7245	0.220A	0.375	T-1	sub midget flg
	7253	0.220A	0.380	T-1	Bi pin
	7636	0.060A	0.600	T-1¼	Bi pin
	7931	0.060A	0.625	T-1¾	Bi pin
	8631	0.060A	0.460	T-1¼	wire term
	8636	0.060A	0.500	T-1¼	special midget flg
	8641	0.060A	0.550	T-1¼	special scr
	8647	0.060A	0.546	T-1¼	knurled scr
	8669	0.060A	0.688	T-1¾	midget scr
	8910	0.220A	0.250	T-1	wire term
1.5v	3225	0.010A	0.380	T-1	Bi pin
	3810	0.015A	0.300	T-¾	micro midget flg

VOLTAGE	LAMP #	CURRENT	MOL	BULB TYPE	BASE
1.5v	3811	0.015A	0.300	T-¾	micro midget flg
	7100	0.015A	0.187	T-¾	wire term
	7101	0.015A	0.187	T-¾	wire term
	7200	0.010A	0.250	T-1	wire term
	7201	0.015A	0.250	T-1	wire term
	7203	0.075A	0.250	T-1	wire term
	7225	0.010A	0.375	T-1	sub midget flg
	7226	0.015A	0.375	T-1	sub midget flg
	7254	0.015A	0.380	T-1	Bi pin
	7255	0.075A	0.380	T-1	Bi pin
	8102	0.075A	0.375	T-1	sub midget flg
2v	48	0.060A	1.190	T-3¼	min scr
	49	0.060A	1.190	T-3¼	min bay
	352	0.060A	0.940	G-3½	min scr
	1809	0.060A	1.190	T-3¼	min bay
2.33v	PR-4	0.270A	1.250	B-3½	S.C. min flg
	233	0.270A	0.940	B-3½	min scr
2.35v	PR-5	0.350A	1.250	B-3½	S.C. min flg
2.38v	PR-2	0.500A	1.250	B-3½	S.C. min flg
2.4v	1491	0.800A	1.750	G-8	D.C. bay
2.46v	245	0.500A	0.940	G-3½	min scr
2.47v	PR-6	0.300A	1.250	B-3½	S.C. min flg
	14	0.300A	0.940	G-3½	min scr
	354	0.300A	0.940	G-3½	min bay
2.5v	10	0.500A	0.940	G-3½	min 2 pin
	35	0.800A	1.250	G-5½	min scr
	41	0.500A	1.190	T-3¼	min scr
	43	0.500A	1.190	T-3¼	min bay
	248	0.800A	1.250	G-5½	min scr
	266	0.350A	0.625	T-1¾	midget groove
	268	0.350A	0.625	T-1¾	midget flg
	326	0.400A	0.546	T-1¼	knurled scr
	329	0.400A	0.550	T-1¼	special scr
	343	0.400A	0.625	T-1¾	midget flg
	368	0.200A	0.625	T-1¾	midget flg
	1492	1.500A	2.000	S-8	D.C. bay
	1769	0.200A	0.688	T-1¾	midget scr
	1783	0.200A	0.520	T-1¾	wire term
	2169	0.350A	0.520	T-1¾	wire term
	7121	0.110A	0.300	T-¾	micro midget flg
	7246	0.015A	0.375	T-1	sub midget flg
	7247	0.100A	0.375	T-1	sub midget flg
	7248	0.320A	0.375	T-1	sub midget flg
	7256	0.015A	0.380	T-1	Bi pin
	7257	0.320A	0.380	T-1	Bi pin
	7307	0.400A	0.600	T-1¼	Bi pin

VOLTAGE	LAMP #	CURRENT	MOL	BULB TYPE	BASE
2.5v	7312	0.350A	0.688	T-1¾	midget scr
	7357	0.400A	0.625	T-1¾	Bi pin
	7732	0.100A	0.380	T-1	Bi pin
	7868	0.350A	0.625	T-1¾	Bi pin
	7968	0.200A	0.625	T-1¾	Bi pin
	8534	0.110A	0.187	T-¾	wire term
	8601	0.015A	0.250	T-1	wire term
	8655	0.400A	0.460	T-1¼	wire term
	8656	0.400A	0.500	T-1¼	special midget flg
	8663	0.400A	0.520	T-1¾	wire term
	8671	0.400A	0.688	T-1¾	midget scr
	8699	0.200A	0.625	T-1¾	midget groove
	8703	0.400A	0.625	T-1¾	midget groove
	8711	0.320A	0.250	T-1	wire term
	8732	0.100A	0.250	T-1	wire term
2.7v	PR-9	0.150A	1.250	B-3½	S.C. min flg
	338	0.060A	0.625	T-1¾	midget flg
	1738	0.060A	0.525	T-1¾	wire term
	7838	0.060A	0.625	T-1¾	Bi pin
	8693	0.060A	0.688	T-1¾	midget scr
	8704	0.060A	0.625	T-1¾	midget groove
3v	323	0.190A	0.546	T-1¼	knurled scr
	324	0.190A	0.460	T-1¼	wire term
	325	0.190A	0.550	T-1¼	special scr
	375	0.015A	0.625	T-1¾	midget flg
	390	0.015A	0.625	T-1¾	midget flg
	679	0.0125A	0.375	T-1	sub midget flg
	2128	0.0125A	0.250	T-1	wire term
	2156	0.030A	0.520	T-1¾	wire term
	2158	0.015A	0.520	T-1¾	wire term
	3229	0.015A	0.380	T-1	Bi pin
	7205	0.015A	0.250	T-1	wire term
	7207	0.060A	0.250	T-1	wire term
	7208	0.120A	0.250	T-1	wire term
	7229	0.015A	0.375	T-1	sub midget flg
	7231	0.060A	0.375	T-1	sub midget flg
	7232	0.120A	0.375	T-1	sub midget flg
	7249	0.008A	0.375	T-1	sub midget flg
	7258	0.008A	0.380	T-1	Bi pin
	7259	0.0125A	0.380	T-1	Bi pin
	7260	0.060A	0.380	T-1	Bi pin
	7261	0.120A	0.380	T-1	Bi pin
	7313	0.150A	0.688	T-1¾	midget scr
	7314	0.030A	0.688	T-1¾	midget scr
	7329	0.030A	0.625	T-1¾	midget flg
	7343	0.030A	0.625	T-1¾	midget groove

VOLTAGE	LAMP #	CURRENT	MOL	BULB TYPE	BASE
3v	7358	0.030A	0.625	T-1¾	Bi pin
	7375	0.015A	0.625	T-1¾	Bi pin
	7637	0.190A	0.600	T-1¼	Bi pin
	8637	0.190A	0.500	T-1¼	special midget flg
	8846	0.008A	0.250	T-1	wire term
3.2v	45	0.350A	1.190	T-3¼	min bay
	1490	0.160A	1.190	T-3¼	min bay
3.57v	PR-3	0.500A	1.250	B-3½	S.C. min flg
3.69v	365	0.500A	0.940	G-3½	min scr
3.7v	PR-7	0.300A	1.250	B-3½	S.C. min flg
	13	0.300A	0.940	G-3½	min scr
	1874	2.750A	1.750	T-5	S.C. bay
3.8v	1438	0.430A	0.940	G-3½	min scr
4v	4A	0.190A	1.688	T-2	telephone slide
	4ES	0.040A	0.700	T-2	wire term
	4ESB	0.040A	0.915	T-2	short slide
	403	0.300A	0.940	G-3½	min scr
	1804	0.060A	1.190	T-3¼	min bay
4.5v	30	0.021A	0.187	T-¾	wire term
	673	0.285A	0.550	T-1¼	special scr
	1021	1.250A	2.250	RP-11	S.C. prefocus
	2171	0.120A	0.520	T-1¾	wire term
	3102	0.022A	0.187	T-¾	wire term
	7304	0.285A	0.546	T-1¼	knurled scr
	7305	0.285A	0.500	T-1¼	special midget flg
	7308	0.285A	0.600	T-1¼	Bi pin
	7315	0.120A	0.688	T-1¾	midget scr
	7331	0.120A	0.625	T-1¾	midget flg
	7345	0.120A	0.625	T-1¾	midget groove
	7359	0.120A	0.625	T-1¾	Bi pin
	8816	0.285A	0.500	T-1¼	wire term
4.75v	PR-13	0.500A	1.250	B-3½	S.C. min flg
4.82v	PR-15	0.500A	1.250	B-3½	S.C. min flg
4.9	27	0.300A	1.060	G-4½	min scr
	407	0.300A	1.060	G-4½	min scr
	408	0.300A	1.060	G-4½	min bay
5v	2	0.021A	0.187	T-¾	wire term
	5ES	0.040A	0.700	T-2	wire term
	5ESB	0.040A	0.915	T-2	short slide
	425	0.500A	1.060	G-4½	min scr
	515	0.115A	0.460	T-1¼	wire term
	580	0.060A	0.460	T-1¼	wire term
	583	0.060A	0.460	T-1¼	wire term
	680	0.060A	0.250	T-1	wire term
	682	0.060A	0.375	T-1	sub midget flg
	683	0.060A	0.250	T-1	wire term
	685	0.060A	0.375	T-1	sub midget flg

VOLTAGE	LAMP #	CURRENT	MOL	BULB TYPE	BASE
5v—————	713	0.075A	0.250	T-1	wire term
	714	0.075A	0.375	T-1	sub midget flg
	715	0.115A	0.250	T-1	wire term
	718	0.115A	0.375	T-1	sub midget flg
	1651	0.600A	2.000	S-8	S.C. bay
	1850	0.090A	1.190	T-3¼	min bay
	2022	0.021A	0.145	T-1	wire term
	2200	0.060A	0.520	T-1¾	wire term
	2203	0.115A	0.520	T-1¾	wire term
	2950	0.017A	0.688	T-1¾	midget scr
	3022	0.021A	0.187	T-1	wire term
	3149	0.060A	0.625	T-1¾	Bi pin
	3150	0.060A	0.625	T-1¾	midget flg
	3151	0.060A	0.525	T-1¾	wire term
	3152	0.060A	0.625	T-1¾	midget groove
	3153	0.060A	0.688	T-1¾	midget scr
	3211	0.115A	0.187	T-1	wire term
	3515	0.115A	0.550	T-1¼	special scr
	3516	0.115A	0.546	T-1¼	knurled scr
	3518	0.115A	0.500	T-1¼	special midget flg
	3580	0.060A	0.550	T-1¼	special scr
	3581	0.060A	0.546	T-1¼	knurled scr
	3582	0.060A	0.500	T-1¼	special midget flg
	3583	0.060A	0.550	T-1¼	special scr
	3584	0.060A	0.546	T-1¼	knurled scr
	3585	0.060A	0.500	T-1¼	special midget flg
	3950	0.017A	0.520	T-1¾	wire term
	6022	0.021A	0.250	T-1	wire term
	6150	0.060A	0.250	T-1	wire term
	6151	0.060A	0.187	T-1	wire term
	6152	0.060A	0.145	T-1	wire term
	6153	0.060A	0.187	T-¾	wire term
	6180	0.060A	0.375	T-1	sub midget flg
	6183	0.060A	0.300	T-¾	micro midget flg
	6211	0.080A	0.187	T-1	wire term
	6212	0.080A	0.145	T-1	wire term
	6801	0.060A	0.187	T-1	wire term
	6802	0.060A	0.145	T-1	wire term
	6803	0.060A	0.187	T-¾	wire term
	6831	0.060A	0.187	T-1	wire term
	6832	0.060A	0.145	T-1	wire term
	6833	0.060A	0.187	T-¾	wire term
	6950	0.017A	0.625	T-1¾	midget groove
	7022	0.021A	0.380	T-1	Bi pin
	7102	0.115A	0.187	T-¾	wire term
	7112	0.021A	0.325	T-¾	Bi pin
	7113	0.060A	0.325	T-¾	Bi pin

VOLTAGE	LAMP #	CURRENT	MOL	BULB TYPE	BASE
5v ————	7114	0.060A	0.325	T-¾	Bi pin
	7115	0.060A	0.325	T-¾	Bi pin
	7116	0.075A	0.325	T-¾	Bi pin
	7117	0.080A	0.325	T-¾	Bi pin
	7118	0.115A	0.325	T-¾	Bi pin
	7119	0.115A	0.325	T-¾	Bi pin
	7122	0.021A	0.300	T-¾	micro midget flg
	7123	0.115A	0.300	T-¾	micro midget flg
	7131	0.075A	0.187	T-1	wire term
	7132	0.075A	0.145	T-1	wire term
	7133	0.075A	0.187	T-¾	wire term
	7151	0.115A	0.187	T-1	wire term
	7152	0.115A	0.145	T-1	wire term
	7153	0.115A	0.187	T-¾	wire term
	7210	0.030A	0.250	T-1	wire term
	7211	0.080A	0.250	T-1	wire term
	7212	0.115A	0.145	T-1	wire term
	7213	0.115A	0.250	T-1	wire term
	7216	0.125A	0.250	T-1	wire term
	7234	0.030A	0.375	T-1	sub midget flg
	7235	0.080A	0.375	T-1	sub midget flg
	7236	0.115A	0.375	T-1	sub midget flg
	7239	0.125A	0.375	T-1	sub midget flg
	7250	0.045A	0.375	T-1	sub midget flg
	7251	0.145A	0.375	T-1	sub midget flg
	7262	0.017A	0.380	T-1	Bi pin
	7263	0.030A	0.380	T-1	Bi pin
	7264	0.045A	0.380	T-1	Bi pin
	7265	0.060A	0.380	T-1	Bi pin
	7266	0.080A	0.380	T-1	Bi pin
	7267	0.115A	0.380	T-1	Bi pin
	7268	0.125A	0.380	T-1	Bi pin
	7269	0.145A	0.380	T-1	Bi pin
	7316	0.060A	0.688	T-1¾	midget scr
	7317	0.190A	0.688	T-1¾	midget scr
	7318	0.060A	0.688	T-1¾	midget scr
	7319	0.115A	0.688	T-1¾	midget scr
	7332	0.060A	0.625	T-1¾	midget flg
	7333	0.060A	0.625	T-1¾	midget flg
	7334	0.190A	0.625	T-1¾	midget flg
	7335	0.115A	0.625	T-1¾	midget flg
	7346	0.060A	0.625	T-1¾	midget groove
	7347	0.060A	0.625	T-1¾	midget groove
	7348	0.115A	0.625	T-1¾	midget groove
	7350	0.190A	0.625	T-1¾	midget groove
	7360	0.060A	0.625	T-1¾	Bi pin
	7361	0.060A	0.625	T-1¾	Bi pin

VOLTAGE	LAMP #	CURRENT	MOL	BULB TYPE	BASE
5v	7362	0.115A	0.625	T-1¾	Bi pin
	7363	0.190A	0.625	T-1¾	Bi pin
	7515	0.115A	0.600	T-1¼	Bi pin
	7538	0.015A	0.325	T-¾	Bi pin
	7580	0.060A	0.600	T-1¼	Bi pin
	7583	0.060A	0.600	T-1¼	Bi pin
	7680	0.060A	0.380	T-1	Bi pin
	7683	0.060A	0.380	T-1	Bi pin
	7714	0.075A	0.380	T-1	Bi pin
	7715	0.115A	0.380	T-1	Bi pin
	7950	0.017A	0.625	T-1¾	Bi pin
	8022	0.021A	0.375	T-1	sub midget flg
	8096	0.145A	0.250	T-1	wire term
	8175	0.125A	0.145	T-1	wire term
	8179	0.125A	0.187	T-1	wire term
	8270	0.115A	0.300	T-¾	micro midget flg
	8383	0.080A	0.300	T-¾	micro midget flg
	8537	0.150A	0.187	T-¾	wire term
	8538	0.015A	0.300	T-¾	micro midget flg
	8587	0.060A	0.300	T-¾	micro midget flg
	8605	0.017A	0.375	T-1	sub midget flg
	8666	0.800A	0.187	T-¾	wire term
	8729	0.045A	0.250	T-1	wire term
	8784	0.190A	0.520	T-1¾	wire term
	8805	0.060A	0.520	T-1¾	wire term
	8828	0.075A	0.300	T-¾	micro midget flg
	8913	0.060A	0.300	T-¾	micro midget flg
	8950	0.017A	0.625	T-1¾	midget flg
5.1v	502	0.150A	1.060	G-4½	min scr
	503	0.150A	1.060	G-4½	min bay
5.4v	1611	1.860A	2.000	S-8	S.C. bay
5.5v	958	2.000A	3.000	G-16½	D.C. bay
	1183	6.250A	2.250	RP-11	S.C. bay
	1184	6.250A	2.250	RP-11	D.C. bay
	1188	6.180A	2.250	RP-11	D.C. bay
5.95v	PR-12	0.500A	1.250	B-3½	S.C. min flg
	1501	6.360A	2.250	RP-11	S.C. prefocus
	1503	6.530A	2.250	RP-11	S.C. prefocus
	2530	6.530A	2.250	RP-11	D.C. prefocus
6v	6A	0.140A	1.688	T-2	telephone slide
	6B	0.290A	1.688	T-2	telephone slide
	6C	0.040A	1.688	T-2	telephone slide
	6ES	0.040A	0.700	T-2	wire term
	6ESB	0.040A	0.915	T-2	short slide
	6PS	0.140A	0.857	T-2	wire term
	6PSB	0.140A	1.115	T-2	short slide
	6MB	0.140A	1.187	T-2	min bay

VOLTAGE	LAMP #	CURRENT	MOL	BULB TYPE	BASE
6v	30	0.450A	1.119	T-3¼	min scr
	60	5.500A	2.380	RP-11	S.C. bay
	316	0.700A	1.190	T-3¼	min bay
	328	0.200A	0.625	T-1¾	midget flg
	337	0.200A	0.625	T-1¾	midget groove
	342	0.040A	0.688	T-1¾	midget scr
	345	0.040A	0.625	T-1¾	midget flg
	371	0.060A	0.550	T-1¼	special scr
	634	0.200A	0.460	T-1¼	wire term
	1025	2.350A	2.250	RP-11	S.C. prefocus
	1482	0.450A	1.060	G-4½	min scr
	1483	0.040A	1.060	G-4½	min scr
	1730	0.040A	0.520	T-1¾	wire term
	1768	0.200A	0.688	T-1¾	midget scr
	1784	0.200A	0.520	T-1¾	wire term
	2114	0.060A	0.460	T-1¼	wire term
	7309	0.060A	0.600	T-1¼	Bi pin
	7328	0.200A	0.625	T-1¾	Bi pin
	7336	0.200A	0.625	T-1¾	midget flg
	7364	0.200A	0.625	T-1¾	Bi pin
	7628	0.200A	0.600	T-1¼	Bi pin
	7660	0.040A	0.600	T-1¼	Bi pin
	7945	0.040A	0.625	T-1¾	Bi pin
	8541	0.060A	0.500	T-1¼	special midget flg
	8543	0.060A	0.546	T-1¼	knurled scr
	8628	0.200A	0.500	T-1¼	special midget flg
	8639	0.200A	0.546	T-1¼	knurled scr
	8645	0.200A	0.550	T-1¼	special scr
	8657	0.040A	0.546	T-1¼	knurled scr
	8660	0.040A	0.500	T-1¼	special midget flg
	8661	0.040A	0.460	T-1¼	wire term
	8662	0.040A	0.550	T-1¼	special scr
	8664	0.200A	0.520	T-1¾	wire term
	8687	0.200A	0.688	T-1¾	midget scr
	8705	0.040A	0.625	T-1¾	midget flg
	8706	0.200A	0.625	T-1¾	midget groove
6.15v	31	0.300A	1.060	G-4½	min scr
	605	0.500A	1.060	G-4½	min scr
	1209	4.100A	2.250	RP-11	S.C. prefocus
	1763	4.100A	2.38	S-11	S.C. prefocus
6.2v	1000	3.870A	2.250	RP-11	D.C. bay
	1133	3.910A	2.250	RP-11	S.C. prefocus
	1134	3.910A	2.250	RP-11	D.C. bay
	1323	4.130A	2.250	RP-11	S.C. prefocus
	1724	4.500A	2.380	S-11	D.C. bay
	2330	4.230A	2.250	RP-11	D.C. prefocus

VOLTAGE	LAMP #	CURRENT	MOL	BULB TYPE	BASE
6.3v	12	0.150A	0.940	G-3½	min 2 pin
	40	0.150A	1.190	T-3¼	min scr
	44	0.250A	1.190	T-3¼	min bay
	46	0.250A	1.190	T-3¼	min scr
	47	0.150A	1.190	T-3¼	min bay
	130	0.150A	0.940	G-3½	min bay
	137	0.250A	0.940	G-3½	min bay
	219	0.250A	0.940	G-3½	min bay
	239	0.360A	1.190	T-3¼	min bay
	240	0.360A	1.190	T-3¼	min bay
	349	0.200A	0.625	T-1¾	midget flg
	350	0.150A	0.625	T-1¾	midget flg
	377	0.075A	0.625	T-1¾	midget flg
	378	0.200A	0.688	T-1¾	midget scr
	379	0.200A	0.625	T-1¾	midget groove
	380	0.040A	0.625	T-1¾	midget flg
	381	0.200A	0.625	T-1¾	midget flg
	398	0.200A	0.625	T-1¾	midget groove
	755	0.150A	1.190	T-3¼	min bay
	1739	0.075A	0.520	T-1¾	wire term
	1775	0.075A	0.688	T-1¾	midget scr
	1810	0.400A	1.190	T-3¼	min bay
	1830	0.150A	1.190	T-3¼	min bay
	1847	0.150A	1.190	T-3¼	min bay
	1855	0.800A	1.380	T-4½	S.C. min bay
	1866	0.250A	1.190	T-3¼	min bay
	2112	0.200A	0.520	T-1¾	wire term
	2180	0.040A	0.520	T-1¾	wire term
	2181	0.200A	0.520	T-1¾	wire term
	7310	0.200A	0.600	T-1¼	Bi pin
	7320	0.040A	0.688	T-1¾	midget scr
	7321	0.150A	0.688	T-1¾	midget scr
	7323	0.200A	0.688	T-1¾	midget scr
	7349	0.200A	0.625	T-1¾	Bi pin
	7351	0.040A	0.625	T-1¾	midget groove
	7352	0.150A	0.625	T-1¾	midget groove
	7368	0.150A	0.625	T-1¾	Bi pin
	7377	0.075A	0.625	T-1¾	Bi pin
	7380	0.040A	0.625	T-1¾	Bi pin
	7381	0.200A	0.625	T-1¾	Bi pin
	8350	0.150A	0.520	T-1¾	wire term
	8551	0.200A	0.500	T-1¼	special midget flg
	8552	0.200A	0.550	T-1¼	special scr
	8553	0.200A	0.546	T-1¼	knurled scr
	8610	0.200A	0.460	T-1¼	wire term
	8708	0.075A	0.625	T-1¾	midget groove

VOLTAGE	LAMP #	CURRENT	MOL	BULB TYPE	BASE
6.4v	6	3.000A	2.000	S-8	D.C. bay
	1129	2.630A	2.000	S-8	S.C. bay
	1130	2.600A	2.000	S-8	D.C. bay
	1154	2.630A	2.000	S-8	D.C. index
	1158	2.630A	2.000	S-8	D.C. bay
	1618	2.800A	2.000	S-8	D.C. bay
6.5v	81	1.020A	1.440	G-6	S.C. bay
	81K	1.020A	1.38	G-6	cand scr
	82	1.050A	1.440	G-6	D.C. bay
	209	1.780A	1.750	B-6	S.C. bay
	210	1.780A	1.750	B-6	D.C. bay
	455	0.500A	1.060	G-4½	min bay
	707	1.880A	1.700	B-6	D.C. bay
	808	1.880A	1.280	B-6	wire term
	1489	2.750A	1.750	T-5	S.C. bay
	1493	2.800A	2.000	S-8	D.C. bay
	1630	2.750A	2.000	S-8	D.C. prefocus
	1811	0.400A	1.190	T-3¼	min bay
	1884	1.540A	1.750	T-5	S.C. bay
6.7v	1619	1.900A	2.000	S-8	S.C. bay
6.8v	87	1.910A	2.000	S-8	S.C. bay
	88	1.900A	2.000	S-8	D.C. bay
	88L	2.000A	2.000	S-8	D.C. bay
7.0v	15	0.400A	1.060	G-4½	min 2 pin
	55	0.410A	1.060	G-4½	min bay
	63	0.630A	1.440	G-6	S.C. bay
	63K	0.630A	1.380	G-6	cand scr
	64	0.630A	1.440	G-6	D.C. bay
	162	0.630A	0.550	G-2	midget scr
	1888	0.500A	1.190	T-3¼	min bay
7.2v	PR-18	0.550A	1.250	B-3½	S.C. min flg
7.5v	50	0.220A	0.940	G-3½	min scr
	51	0.220A	0.940	G-3½	min bay
8v	8	2.200A	2.000	S-8	D.C. bay
	8A	0.900A	1.688	T-2	telephone slide
	426	0.250A	1.060	G-4½	min scr
	1648	2.000A	2.000	S-8	D.C. bay
8.63v	PR-20	0.500A	1.250	B-3½	S.C. min flg
10v	10A	0.110A	1.688	T-2	telephone slide
	10B	0.250A	1.688	T-2	telephone slide
	10AS	0.010A	0.700	T-2	wire term
	10ASB	0.010A	0.915	T-2	short slide
	10CS	0.017A	0.700	T-2	wire term
	10CSB	0.017A	0.915	T-2	short slide
	10ES	0.040A	0.700	T-2	wire term
	10ESB	0.040A	0.915	T-2	short slide

VOLTAGE	LAMP #	CURRENT	MOL	BULB TYPE	BASE
10v	240	1.020A	0.940	T-3	wire term
	344	0.014A	0.625	T-1¾	midget flg
	367	0.040A	0.625	T-1¾	midget flg
	389	0.040A	0.688	T-1¾	midget scr
	397	0.040A	0.625	T-1¾	midget groove
	709	0.014A	0.625	T-1¾	midget groove
	755	0.450A	0.550	T-3	wire term
	1869	0.014A	0.520	T-1¾	wire term
	2107	0.040A	0.520	T-1¾	wire term
	7218	0.027A	0.250	T-1	wire term
	7240	0.027A	0.375	T-1	sub midget flg
	7311	0.014A	0.600	T-1¼	Bi pin
	7344	0.014A	0.625	T-1¾	Bi pin
	7367	0.040A	0.625	T-1¾	Bi pin
	8095	0.027A	0.380	T-1	Bi pin
	8606	0.014A	0.550	T-1¼	special scr
	8607	0.014A	0.546	T-1¼	knurled scr
	8608	0.014A	0.460	T-1¼	wire term
	8609	0.014A	0.500	T-1¼	special midget flg
	8691	0.014A	0.688	T-1¾	midget scr
11v	7325	0.022A	0.688	T-1¾	midget scr
	7338	0.022A	0.625	T-1¾	midget flg
	7353	0.022A	0.625	T-1¾	midget groove
	7369	0.022A	0.625	T-1¾	Bi pin
	8946	0.022A	0.520	T-1¾	wire term
12v	12A	0.100A	1.688	T-2	telephone slide
	12B	0.350A	1.688	T-2	telephone slide
	12C	0.170A	1.688	T-2	telephone slide
	12ES	0.040A	0.700	T-2	wire term
	12ESB	0.040A	0.915	T-2	short slide
	12PS	0.070A	0.857	T-2	wire term
	12PSB	0.170A	1.115	T-2	short slide
	12MB	0.170A	1.187	T-2	min bay
	32	0.060A	0.375	T-1	sub midget flg
	394	0.040A	0.625	T-1¾	midget flg
	1446	0.200A	0.940	G-3½	min scr
	1471	0.260A	1.380	G-6	cand scr
	2174	0.040A	0.520	T-1¾	wire term
	7219	0.060A	0.250	T-1	wire term
	7326	0.040A	0.688	T-1¾	midget scr
	7354	0.040A	0.625	T-1¾	midget groove
	7371	0.040A	0.625	T-1¾	Bi pin
	8097	0.060A	0.380	T-1	Bi pin
12.5v	PR-16	0.250A	1.250	B-3½	S.C. min flg
	428	0.250A	1.060	G-4½	min scr
	1124	1.920A	2.250	RP-11	D.C. bay

VOLTAGE	LAMP #	CURRENT	MOL	BULB TYPE	BASE
12.5v	1143	2.000A	2.250	RP-11	S.C. bay
	1144	2.000A	2.250	RP-11	D.C. bay
	1195	3.000A	2.250	RP-11	S.C. bay
	1196	3.000A	2.250	RP-11	D.C. bay
	1507	3.000A	2.250	RP-11	S.C. prefocus
	2336	2.230A	2.250	RP-11	D.C. prefocus
12.8v	93	1.040A	2.000	S-8	S.C. bay
	94	1.000A	2.000	S-8	D.C. bay
	1003	0.940A	1.750	B-6	S.C. bay
	1004	0.940A	1.750	B-6	S.C. bay
	1005	1.260A	2.000	S-8	D.C. bay
	1011	2.080A	2.250	RP-11	S.C. prefocus
	1016	1.340A	2.000	S-8	D.C. index
	1026	2.180A	2.250	RP-11	D.C. prefocus
	1034	1.800A	2.000	S-8	D.C. index
	1044	1.850A	2.250	RP-11	D.C. prefocus
	1073	1.800A	2.000	S-8	S.C. bay
	1076	1.800A	2.000	S-8	D.C. bay
	1141	1.440A	2.000	S-8	S.C. bay
	1142	1.440A	2.000	S-8	D.C. bay
	1156	2.100A	2.000	S-8	S.C. bay
	1157	2.100A	2.000	S-8	D.C. index
	1176	1.340A	2.000	S-8	D.C. bay
	1327	2.080A	2.250	RP-11	S.C. prefocus
13v	78	0.370A	1.380	G-5	D.C. bay
	89	0.580A	1.380	G-6	S.C. bay
	89K	0.580A	1.380	G-6	cand scr
	90	0.580A	1.440	G-6	D.C. bay
	1816	0.330A	1.190	T-3¼	min bay
13.5v	67	0.590A	1.440	G-6	S.C. bay
	67K	0.410A	1.380	G-6	cand scr
	67M	0.410A	1.310	G-6	min scr
	68	0.590A	1.440	G-6	D.C. bay
	1155	0.590A	1.440	G-6	S.C. bay
14v	57	0.240A	1.060	G-4½	min bay
	57X	0.240A	1.060	G-4½	min bay
	163	0.065A	0.550	G-2	midget scr
	257	0.270A	1.060	G-4½	min bay
	258	0.270A	1.060	G-4½	min scr
	330	0.080A	0.625	T-1¾	midget flg
	336	0.080A	0.625	T-1¾	midget groove
	363	0.200A	0.940	G-3½	min bay
	373	0.080A	0.688	T-1¾	midget scr
	382	0.080A	0.625	T-1¾	midget scr
	386	0.080A	0.625	T-1¾	midget groove
	393	0.100A	0.625	T-1¾	midget groove

VOLTAGE	LAMP #	CURRENT	MOL	BULB TYPE	BASE
14v	431	0.250A	1.060	G-4½	min bay
	631	0.630A	1.440	G-6	S.C. bay
	756	0.080A	1.190	T-3¼	min bay
	1074	0.510A	2.250	RP-11	D.C. prefocus
	1247	0.480A	1.440	G-6	S.C. bay
	1449	0.200A	0.940	G-3½	min scr
	1474	0.170A	1.250	T-3	min scr
	1479	0.170A	1.500	T-4	cand scr
	1487	0.200A	1.190	T-3¼	min scr
	1488	0.150A	1.190	T-3¼	min bay
	1705	0.080A	0.520	T-1¾	wire term
	1815	0.200A	1.190	T-3¼	min bay
	1889	0.270A	1.190	T-3¼	min bay
	1891	0.240A	1.190	T-3¼	min bay
	1893	0.330A	1.190	T-3¼	min bay
	1895	0.270A	1.060	G-4½	min bay
	2162	0.100A	0.520	T-1¾	wire term
	2182	0.080A	0.520	T-1¾	wire term
	7330	0.080A	0.625	T-1¾	Bi pin
	7646	0.080A	0.600	T-1¼	Bi pin
	7373	0.100A	0.625	T-1¾	Bi pin
	7382	0.080A	0.625	T-1¾	Bi pin
	8098	0.065A	0.380	T-1	Bi pin
	8111	0.065A	0.250	T-1	wire term
	8112	0.065A	0.375	T-1	sub midget flg
	8162	0.100A	0.688	T-1¾	midget scr
	8362	0.080A	0.688	T-1¾	midget scr
	8640	0.080A	0.460	T-1¼	wire term
	8644	0.080A	0.550	T-1¼	special scr
	8646	0.080A	0.500	T-1¼	special midget flg
	8654	0.080A	0:546	T-1¼	knurled scr
	8918	0.100A	0.625	T-1¾	midget flg
14.4v	19	0.100A	0.940	G-3½	min 2 pin
	52	0.100A	0.940	G-3½	min scr
	53	0.120A	0.940	G-3½	min bay
	53X	0.120A	0.940	G-3½	min bay
	1813	0.100A	1.190	T-3¼	min bay
	1892	0.120A	1.190	T-3¼	min bay
16v	16A	0.100A	1.688	T-2	telephone slide
	16B	0.290A	1.688	T-2	telephone slide
	16CS	0.017A	0.700	T-2	wire term
	16CSB	0.017A	0.915	T-2	short slide
	16ES	0.040A	0.700	T-2	wire term
	16ESB	0.040A	0.915	T-2	short-slide
18v	18A	0.040A	1.688	T-2	telephone slide
	18B	0.100A	1.688	T-2	telephone slide

VOLTAGE	LAMP #	CURRENT	MOL	BULB TYPE	BASE
18v————	18ES	0.040A	0.700	T-2	wire term
	18ESB	0.040A	0.915	T-2	short slide
	346	0.040A	0.625	T-1¾	midget groove
	370	0.040A	0.625	T-1¾	midget flg
	432	0.250A	1.060	G-4½	min scr
	433	0.250A	1.060	G-4½	min bay
	1445	0.150A	0.940	G-3½	min bay
	1447	0.150A	0.940	G-3½	min scr
	1456	0.250A	1.190	G-5	min bay
	1476	0.170A	1.250	T-3	min scr
	1480	0.170A	1.500	T-4	cand scr
	1826	0.150A	1.190	T-3¼	min bay
	2102	0.040A	0.520	T-1¾	wire term
	7220	0.026A	0.250	T-1	wire term
	7241	0.026A	0.375	T-1	sub midget flg
	7370	0.040A	0.625	T-1¾	Bi pin
	8099	0.026A	0.380	T-1	Bi pin
	8536	0.040A	0.688	T-1¾	midget scr
20v————	20C	0.034A	1.688	T-2	telephone slide
	1458	0.250A	1.190	G-5	min bay
22v————	71	0.180A	1.440	G-6	S.C. bay
	71K	0.180A	1.380	G-6	cand scr
	71M	0.180A	1.310	G-6	min scr
	72	0.180A	1.440	G-6	D.C. bay
	457	0.040A	0.625	T-1¾	midget groove
	459	0.040A	0.625	T-1¾	midget flg
	1464	0.250A	1.190	G-5	min bay
	7459	0.040A	0.625	T-1¾	Bi pin
	8425	0.040A	0.520	T-1¾	wire term
	8437	0.040A	0.688	T-1¾	midget scr
24v————	24A	0.030A	1.688	T-2	telephone slide
	24B	0.040A	1.688	T-2	telephone slide
	24C	0.072A	1.688	T-2	telephone slide
	24CS	0.017A	0.700	T-2	wire term
	24CSB	0.017A	0.915	T-2	short slide
	24D	0.100A	1.688	T-2	telephone slide
	24E	0.035A	1.688	T-2	telephone slide
	24ES	0.040A	0.700	T-2	wire term
	24ESB	0.040A	0.915	T-2	short slide
	24F	0.090A	1.688	T-2	telephone slide
	24MB	0.073A	1.187	T-2	min bay
	24PS	0.073A	0.857	T-2	wire term
	24PSB	0.073A	1.115	T-2	short slide
	24X	0.035A	1.688	T-2	telephone slide
	509	0.180A	1.440	G-6	S.C. bay
	509K	0.180A	1.380	G-6	cand scr

VOLTAGE	LAMP #	CURRENT	MOL	BULB TYPE	BASE
24v	530	0.170A	1.380	G-6	cand scr
	1448	0.035A	0.940	G-3½	min scr
	1477	0.170A	1.250	T-3	min scr
	1818	0.170A	1.190	T-3¼	·min bay
	1841	0.170A	1.500	T4	cand scr
	2176	0.050A	0.520	T-1¾	wire term
	7001	0.050A	0.625	T-1¾	Bi pin
	8176	0.050A	0.625	T-1¾	midget flg
	8177	0.050A	0.625	T-1¾	midget groove
	8178	0.050A	0.688	T-1¾	midget scr
25.6v	1271	1.340A	2.250	RP-11	S.C. prefocus
26v	1047	2.700A	2.250	RP-11	S.C. bay
28v	28ES	0.040A	0.700	T-2	wire term
	28ESB	0.040A	0.915	T-2	short slide
	28PS	0.040A	0.857	T-2	wire term
	28PSB	0.040A	1.115	T-2	short slide
	28MB	0.040A	1.187	T-2	min bay
	301	0.170A	1.380	G-5	S.C. bay
	302	0.170A	1.380	G-5	D.C. bay
	303	0.300A	1.440	G-6	S.C. bay
	304	0.300A	1.440	G-6	D.C. bay
	305	0.560A	2.000	S-8	S.C. bay
	306	0.510A	2.000	S-8	D.C. bay
	307	0.660A	2.000	S-8	S.C. bay
	308	0.670A	2.000	S-8	D.C. bay
	309	0.900A	2.380	S-11	S.C. bay
	310	0.920A	2.380	S-11	D.C. bay
	311	1.290A	2.380	S-11	S.C. bay
	313	0.170A	1.190	T-3¼	min bay
	315	0.900A	2.000	S-8	S.C. bay
	327	0.040A	0.625	T-1¾	midget flg
	334	0.040A	0.625	T-1¾	midget groove
	335	0.040A	0.688	T-1¾	midget scr
	336	0.170A	0.940	G-3½	min bay
	376	0.060A	0.625	T-1¾	midget flg
	385	0.040A	0.625	T-1¾	midget flg
	387	0.040A	0.625	T-1¾	midget flg
	388	0.040A	0.625	T-1¾	midget groove
	399	0.040A	0.688	T-1¾	midget scr
	623	0.370A	1.440	G-6	S.C. bay
	1203	0.700A	2.000	S-8	S.C. bay
	1204	0.700A	2.000	S-8	D.C. bay
	1251	0.230A	1.440	G-6	S.C. bay
	1252	0.230A	1.440	G-6	D.C. bay
	1309	0.520A	1.750	B-6	S.C. bay
	1495	0.300A	1.380	T-4½	S.C. min bay

VOLTAGE	LAMP #	CURRENT	MOL	BULB TYPE	BASE
28v	1666	0.680A	2.000	S-8	D.C. bay
	1683	1.020A	2.000	S-8	S.C. bay
	1691	0.610A	2.000	S-8	S.C. bay
	1692	0.610A	2.000	S-8	D.C. bay
	1696	0.420A	2.000	S-8	D.C. bay
	1764	0.040A	0.520	T-1¾	wire term
	1819	0.040A	1.190	T-3¼	min bay
	1820	0.100A	1.190	T-3¼	min bay
	1821	0.170A	1.190	T-3¼	min scr
	1829	0.070A	1.190	T-3¼	min bay
	1864	0.170A	1.190	T-3¼	min bay
	2185	0.040A	0.520	T-1¼	wire term
	2187	0.040A	0.520	T-1¾	wire term
	6838	0.024A	0.250	T-1	wire term
	6839	0.024A	0.375	T-1	sub midget flg
	7327	0.040A	0.625	T-1¾	Bi pin
	7341	0.065A	0.625	T-1¾	midget flg
	7355	0.040A	0.625	T-1¾	midget groove
	7356	0.065A	0.625	T-1¾	midget groove
	7374	0.040A	0.625	T-1¾	Bi pin
	7376	0.065A	0.615	T-1¾	Bi pin
	7387	0.040A	0.625	T-1¾	Bi pin
	7632	0.040A	0.600	T-1¼	Bi pin
	7839	0.024A	0.380	T-1	Bi pin
	7876	0.060A	0.625	T-1¾	Bi pin
	8361	0.065A	0.520	T-1¾	wire term
	8369	0.065A	0.688	T-1¾	midget scr
	8623	0.040A	0.546	T-1¼	knurled scr
	8627	0.040A	0.460	T-1¼	wire term
	8632	0.040A	0.500	T-1¼	special midget flg
	8635	0.040A	0.550	T-1¼	special scr
32v	1054	0.740A	2.250	RP-11	D.C. bay
	1056	1.150A	2.250	RP-11	D.C. bay
	1224K	0.160A	1.380	G-6	cand scr
	1226	0.210A	1.750	G-8	D.C. bay
	1230	0.530A	2.250	RP-11	D.C. bay
	1238	3.500A	3.000	G-16½	D.C. bay
	1240	3.500A	3.000	G-16½	D.C. prefocus
34v	1223	0.160A	1.440	G-6	S.C. bay
	1224	0.160A	1.440	G-6	D.C. bay
	1228	0.450A	2.000	S-8	D.C. bay
44v	44A	0.072A	1.688	T-2	telephone slide
	109	0.170A	2.000	S-8	S.C. bay
	110	0.170A	2.000	S-8	D.C. bay
	1150	0.470A	2.250	RP-11	D.C. bay
45v	536	0.200A	1.440	G-6	D.C. bay

5.9   RADIO FREQUENCY CABLE SPECIFICS

Impedance	RG/U#	Diameters Inch	Millimeter	Maximum Operating Voltage
25 ohms————191		1.460	37.084	15,000 vrms
35 ohms————83		0.405	10.287	—
48 ohms————25		0.565	14.351	8,000 V peak
	25A	0.505	12.827	10,000 V peak
	25B	0.750	19.050	15,000 V peak
	26	0.525	13.335	8,000 V peak
	26A	0.505	12.827	10,000 V peak
	27	0.675	17.145	15,000 V peak
	27A	0.670	17.018	15,000 V peak
	28	0.805	20.447	15,000 V peak
	28B	0.750	19.050	15,000 V peak
	64	0.495	12.573	8,000 V peak
	64A	0.475	12.065	10,000 V peak
	77	0.414	10.516	—
	78	0.385	9.779	—
	88	0.515	13.081	10,000 Vrms
50 ohms————5A		0.328	8.331	—
	9B	0.420	10.668	—
	55B	0.206	5.232	1,900 Vrms
	58A	0.195	4.953	—
	58C	0.195	4.959	1,900 Vrms
	60	0.425	10.795	—
	87	0.425	10.795	—
	115	0.375	9.525	—
	117	0.730	18.542	—
	119	0.470	11.938	—
	122	0.160	4.064	—
	126	0.290	7.366	—
	141	0.195	4.953	—
	141A	0.190	4.826	—
	142	0.206	5.232	—
	142B	0.195	4.953	—
	143	0.325	8.255	—
	156	0.540	13.716	10,000 Vrms
	157	0.725	18.415	15,000 Vrms
	158	0.725	18.415	15,000 Vrms
	174	0.100	2.540	—
	178B	0.075	1.905	1,000 Vrms
	179B	0.105	2.667	1,200 Vrms
	188A	0.102	2.591	—
	190	0.700	17.780	15,000 Vrms

Impedance	RG/U#	Diameters Inch	Millimeter	Maximum Operating Voltage
50 ohms———	196A	0.080	2.032	1,000 Vrms
	211A	0.730	18.542	7,000 Vrms
	212	0.332	8.433	3,000 Vrms
	213	0.405	10.287	5,000 Vrms
	214	0.425	10.795	5,000 Vrms
	215	0.405	10.287	5,000 Vrms
	217	0.545	13.843	7,000 Vrms
	218	0.870	22.098	11,000 Vrms
	219	0.603	15.316	11,000 Vrms
	220	1.120	28.448	14,000 Vrms
	221	1.195	30.353	14,000 Vrms
	223	0.216	5.486	1,900 Vrms
	224	0.615	15.621	7,000 Vrms
	225	0.430	10.922	5,000 Vrms
	226	0.500	12.700	7,000 Vrms
	227	0.490	12.446	5,000 Vrms
	228A	0.795	20.193	7,000 Vrms
	301	0.245	6.223	3,000 Vrms
	303	0.170	4.318	1,900 Vrms
	304	0.280	7.112	3,000 Vrms
	316	0.102	2.591	1,200 Vrms
51 ohms———	9	0.420	10.668	4,000 Vrms
	9A	0.420	10.668	4,000 Vrms
	33	0.470	11.938	—
52 ohms———	8	0.405	10.287	4,000 Vrms
	8A	0.405	10.287	—
	10	0.475	12.065	4,000 Vrms
	14	0.545	13.843	5,500 Vrms
	16	0.630	16.002	—
	17	0.870	22.098	11,000 Vrms
	18	0.945	24.003	11,000 Vrms
	19	1.120	28.448	14,000 Vrms
	20	1.195	30.353	14,000 Vrms
	74	0.615	15.621	5,500 Vrms
	212	0.332	8.433	—
	213	0.405	10.287	—
	217	0.545	13.843	—
	218	0.870	22.098	—
	219	0.945	24.003	—
	220	1.120	28.448	—
52.5 ohms———	5	0.332	8.433	3,000 Vrms
53 ohms———	21	0.332	8.433	2,700 Vrms
53.5 ohms———	55	0.206	5.232	1,900 Vrms
	55B	0.206	5.232	—

Impedance	RG/U#	Diameters Inch	Millimeter	Maximum Operating Voltage
53.5 ohms———————58		0.195	4.953	1,900 Vrms
58 ohms———————54A		0.245	6.223	3,000 Vrms
67.5 ohms———————41		0.425	10.795	3,000 Vrms
69.0 ohms———————36		1.180	29.972	—
71 ohms———————34		0.625	15.875	5,200 Vrms
	35	0.945	24.003	10,000 Vrms
72 ohms———————144		0.395	10.033	—
73 ohms———————59		0.242	6.147	2,300 Vrms
	140	0.242	6.147	—
74 ohms———————13		0.420	10.668	4,000 Vrms
	216	0.425	10.795	—
75 ohms——————— 6A		0.332	8.433	2,700 Vrms
	11	0.405	10.287	4,000 Vrms
	11A	0.412	10.465	5,000 Vrms
	12A	0.475	12.065	5,000 Vrms
	34B	0.630	16.002	6,500 Vrms
	35B	0.945	25.003	10,000 Vrms
	59B	0.242	6.147	2,300 Vrms
	84A	1.000	25.400	10,000 Vrms
	85A	1.565	39.751	10,000 Vrms
	101	0.588	14.935	—
	164	0.870	22.098	10,000 Vrms
	179A	0.100	2.540	—
	187	0.110	2.794	1,200 Vrms
	216	0.425	10.795	5,000 Vrms
	302	0.206	5.232	2,300 Vrms
	307	0.270	6.858	400 Vrms
	144	0.410	10.414	5,000 Vrms
76 ohms——————— 6		0.332	8.433	2,700 Vrms
	6A	0.332	8.433	—
	15	0.545	13.843	5,000 Vrms
	108	0.245	6.223	—
78 ohms———————42		0.342	8.687	2,700 Vrms
93 ohms———————62A		0.242	6.147	750 Vrms
	62B	0.242	6.147	750 Vrms
	71	0.250	6.350	750 Vrms
	71B	0.250	6.350	750 Vrms
95 ohms——————— 22		0.405	10.287	1,000 Vrms
	22B	0.420	10.668	1,000 Vrms
	57	0.625	15.875	3,000 Vrms
	57A	0.625	15.875	3,000 Vrms
	65	0.405	10.287	1,000 Vrms
	111A	0.490	12.446	1,000 Vrms
	130	0.625	15.875	8,000 Vrms

Impedance	RG/U#	Diameters Inch	Millimeter	Maximum Operating Voltage
95 ohms————131	131	0.710	18.034	8,000 Vrms
	180B	0.145	3.683	1,500 Vrms
	195A	0.155	3.937	1,500 Vrms
125 ohms————23	23	0.945	24.003	3,000 Vrms
	63B	0.405	10.287	1,000 Vrms
	79B	0.475	12.065	1,000 Vrms
	181	0.640	16.256	3,500 Vrms
	89	0.632	16.053	—
140 ohms————102	102	1.088	27.635	—
185 ohms————114	114	0.405	10.287	—

5.10   VHF AND UHF TV CHANNEL SPECTRUM

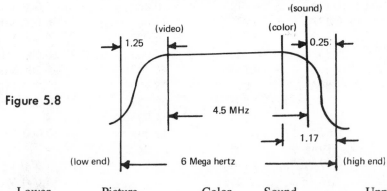

Figure 5.8

Channel Number	Lower Frequency (MHz)	Picture Carrier (AM) (MHz)	Color Carrier (MHz)	Sound Carrier (FM) (MHz)	Upper Frequency (MHz)
(Low VHF)					
2	54	55.25	58.85	59.75	60
3	60	61.25	64.83	65.75	66
4	66	67.25	70.83	71.75	72

GOVERNMENTAL, AERONAUTICAL, AND ASTRONOMY ASSIGNMENTS

5	76	77.25	80.83	81.75	82
6	82	83.25	86.83	87.75	88

FM, GOVERNMENTAL, AERONAUTICAL, TELEMETERING, INDUSTRIAL, AMATEUR AND MISC. ASSIGNMENTS

(High VHF)					
7	174	175.25	178.83	179.75	180
8	180	181.25	184.83	185.75	186
9	186	187.25	190.83	191.75	192

Channel Number	Lower Frequency (MHz)	Picture Carrier (AM) (MHz)	Color Carrier (MHz)	Sound Carrier (FM) (MHz)	Upper Frequency (MHz)
10	192	193.25	196.83	197.75	198
11	198	199.25	202.83	203.75	204
12	204	205.25	208.83	209.75	210
13	210	211.25	214.83	215.75	216

GOVERNMENTAL, TELEMETERING, AMATEUR, AERONAUTICAL, INDUSTRIAL AND MISC. ASSIGNMENTS

(UHF)

Channel	Lower	Picture Carrier	Color	Sound	Upper
14	470	471.25	474.83	475.75	476
15	476	477.25	480.83	481.75	482
16	482	483.25	486.83	487.75	488
17	488	489.75	492.83	493.75	494
18	494	495.25	498.83	499.75	500
19	500	501.25	504.83	505.75	506
20	506	507.25	510.83	511.75	512
21	512	513.25	516.83	517.75	518
22	518	519.25	522.83	523.75	524
23	524	525.25	528.83	529.75	530
24	530	531.25	534.83	535.75	536
25	536	537.25	540.83	541.75	542
26	542	543.25	546.83	547.75	548
27	548	549.25	552.83	553.75	554
28	554	555.25	558.83	559.75	560
29	560	561.25	564.83	565.75	566
30	566	567.25	570.83	571.75	572
31	572	573.25	576.83	577.75	578
32	578	579.25	582.83	583.75	584
33	584	585.25	587.83	589.75	590
34	590	591.25	594.83	595.75	596
35	596	597.25	600.83	601.75	602
36	602	603.25	606.83	607.75	608
37	608	609.25	612.83	613.75	614
38	614	615.25	618.83	619.75	620
39	620	621.25	624.83	625.75	626
40	626	627.25	630.83	631.75	632
41	632	633.25	636.83	637.75	638
42	638	639.25	642.83	643.75	644
43	644	645.25	648.83	649.75	650
44	650	651.25	654.83	655.75	656
45	656	657.25	660.83	661.75	662
46	662	663.25	666.83	667.75	668

Channel Number	Lower Frequency (MHz)	Picture Carrier (AM) (MHz)	Color Carrier (MHz)	Sound Carrier (FM) (MHz)	Upper Frequency (MHz)
47	668	669.25	672.83	673.75	674
48	674	675.25	678.83	679.75	680
49	680	681.25	684.83	685.75	686
50	686	687.25	690.83	691.75	692
51	692	693.25	696.83	697.75	698
52	698	699.25	702.75	703.75	704
53	704	705.25	707.83	709.75	710
54	710	711.25	714.83	715.75	716
55	716	717.25	720.83	721.75	722
56	722	723.25	726.83	727.75	728
57	728	729.25	732.83	733.75	734
58	734	735.25	738.83	739.75	740
59	740	741.25	744.83	745.75	746
60	746	747.25	780.83	751.75	752
61	752	753.25	756.83	757.75	758
62	758	759.25	762.83	763.75	764
63	764	765.25	768.83	769.75	770
64	770	771.25	774.83	775.75	776
65	776	777.25	780.83	781.75	782
66	782	783.25	786.83	787.75	788
67	788	789.25	792.83	793.75	794
68	794	795.25	798.83	799.75	800
69	800	801.25	804.83	805.75	806
70	806	807.25	810.83	811.75	812
71	812	813.25	816.83	817.75	818
72	818	819.25	822.83	823.75	824
73	824	825.25	828.83	829.75	830
74	830	831.25	834.83	835.75	836
75	836	837.25	840.83	841.75	842
76	842	843.25	846.83	847.75	848
77	848	849.25	852.83	853.75	854
78	854	855.25	858.83	859.75	860
79	860	861.25	864.83	865.75	866
80	866	867.25	870.83	871.75	872
81	872	873.25	876.83	877.75	878
82	878	879.25	882.83	883.75	884
83	884	885.25	888.83	889.75	890

CHAPTER 6

Alphabetized Conversion Factors for Technical Applications

Not too long ago the word electronics meant only radio and television, but in this modern day electronics is employed one way or another in every imaginable facet of life. The data for calculations, normally used by only a few highly trained individuals, is now the language used by most electronics professional personnel.

It is the intent of this chapter to share with you the "working tool information" for converting from one unit to another simply by multiplying by its conversion factor.

Over 800 units have been alphabetized. If you were working with centimeters and wished to convert into meters, look under the c's for centimeter and then for the conversion factor required to convert into meters.

	TO CONVERT	INTO	MULTIPLY BY
(A)	Abamperes	Amperes	10
	Abamperes	Statamperes	3×10^{10}
	Abamperes/sq. cm.	Amperes/sq. inch	64.52
	Abampere-turns	Ampere-turns	10
	Abampere-turns	Gilberts	12.57
	Abampere-turns/cm.	Ampere-turns/inch	25.40
	Abcoulombs	Coulombs	10
	Abcoulombs	Statcoulombs	3×10^{10}
	Abcoulombs/sq. cm.	Coulombs/sq inch	64.52
	Abfarads	Farads	1×10^9
	Abfarads	Microfarads	1×10^{15}
	Abfarads	Statfarads	9×10^{20}
	Abhenries	Henries	1×10^{-9}
	Abhenries	Millihenries	1×10^{-6}
	Abhenries	Stathenries	$^1/_9 \times 10^{-20}$
	Abohms/cm. cube	Microhms/cm. cube	1×10^{-3}
	Abohms/cm. cube	Ohms/mil foot	6.015×10^{-3}
	Abvolts	Statvolts	$^1/_3 \times 10^{-10}$
	Abvolts	Volts	1×10^{-8}
	Acres	Square Feet	43,560.00
	Acres	Square meters	4,047.00
	Acres	Square Miles	1.562×10^{-3}
	Acres	Square Yards	4,840.00

TO CONVERT	INTO	MULTIPLY BY
Acre-feet	Cubic-feet	43,560.00
Acre-feet	Gallons	3.259×10^5
Amperes	Abamperes	$^1/_{10}$
Amperes	Microamperes	1×10^6
Amperes	Milliamperes	1×10^3
Amperes	Statamperes	3×10^9
Amperes/sq. cm.	Amperes/sq. inch	6.452
Amperes/sq. cm.	Amperes/sq. meter	1×10^4
Amperes/sq. inch	Abamperes/sq.cm.	0.0155
Amperes/sq. inch	Amperes/sq. meter	1,550.00
Amperes/sq. inch	Amperes/sq.cm.	0.1550
Amperes/sq. inch	Statamperes/sq. cm.	4.650×10^8
Amperes/sq. meter	Amperes/sq. cm.	1×10^{-4}
Amperes/sq. meter	Amperes/sq. inch	6.452×10^{-4}
Ampere-hours	Coulombs	3,600.00
Ampere-hours	Faradays	0.03731
Ampere-turns	Abampere-turns	$^1/_{10}$
Ampere-turns	Gilberts	1.257
Ampere-turns/cm.	Ampere-turns/inch	2.540
Ampere-turns/cm.	Ampere-turns/meter	100.00
Ampere-turns/cm.	Gilberts/cm.	1.257
Ampere-turns/inch	Abampere-turns/cm.	0.03937
Ampere-turns/inch	Ampere-turns/cm.	0.3937
Ampere-turns/inch	Gilberts/cm.	0.4950
Ampere-turns/inch	Ampere-turns/meter	39.37
Ampere-turns/meter	Ampere-turns/cm.	0.01
Ampere-turns/meter	Ampere-turns/inch	0.0254
Ampere-turns/meter	Gilberts/cm.	0.01257
Angstrum units	Inches	3.937×10^{-8}
Angstrum units	Meters	1×10^{-10}
Atmospheres	Cms of mercury	76.00
Atmospheres	Feet of water (@ 4° c)	33.90
Atmospheres	inches of mercury (@ 0° c)	29.92
Atmospheres	Kgs/sq. cm.	1.033
Atmospheres	Kgs/sq. meter	10,332.00
Atmospheres	Pounds/sq. inch	14.70
Atmospheres	Tons/sq. foot	1.058
(B) Barrels (oil)	Gallons (oil)	42.00
Barns	Sq. cm.	1×10^{-24}
Bars	Atmospheres	9.870×10^{-7}
Bars	Dynes/sq. cm.	1×10^6
Bars	Kgs/sq. meter	1.020×10^4
Bars	Pounds/sq. ft.	2.089×10^3
Bars	Pounds/sq. inch	14.50

TO CONVERT	INTO	MULTIPLY BY
Btu	Ergs	1.055×10^{10}
Btu	Foot-pounds	778.30
Btu	Gram-calories	252.00
Btu	Horsepower-hrs.	3.927×10^{-4}
Btu	Joules	1054.00
Btu	Kilogram-calories	0.2520
Btu	Kilogram-meters	107.5
Btu	Kilowatt-hrs.	2.928×10^{-4}
Btu/hour	Ft-lbs/sec	0.2162
Btu/hour	Gram-cal/sec	0.070
Btu/hour	Horsepower-hrs.	3.929×10^{-4}
Btu/hour	Watts	0.2931
Btu/minute	Ft-lbs/sec	12.96
Btu/minute	Horsepower	0.02356
Btu/minute	Kilowatts	0.01757
Btu/minute	Watts	17.57
Btu/sq.ft./minute	Watts/sq. inch	0.1221
Bushels	Cubic foot	1.2445
Bushels	Cubic inch	2,150.400
Bushels	Cubic meters	0.03524
Bushels	Liters	35.24
Bushels	Pecks	4.0
Bushels	Pints (dry)	64.0
Bushels	Quarts (dry)	32.0
(C) Centares	Sq. meters	1.0
Centigrade	Fahrenheit	$(c° \times 9/5) + 32$
Centigrade	Celsius	1.0
Centigrade	Kelvin	$(C° + 273.1)$
Centigrams	Grams	0.01
Centiliters	Liters	0.01
Centimeters	Feet	3.281×10^{-2}
Centimeters	Inches	0.3937
Centimeters	Kilometers	1×10^{-5}
Centimeters	Meters	0.01
Centimeters	Miles	6.214×10^{-6}
Centimeters	Millimeters	10.0
Centimeters	Mils	393.7
Centimeters	Yards	1.094×10^{-2}
Centimeter-dynes	Meter-kilograms	1.020×10^{-8}
Centimeter-dynes	Pound-feet	7.376×10^{-8}
Centimeter-grams	Centimeter-dynes	980.7
Centimeter-grams	Meter-kilograms	1×10^{-5}
Centimeter-grams	Pound-feet	7.233×10^{-5}
Centimeters of mercury	Atmospheres	0.01316
Centimeters of mercury	Feet of water	0.4461
Centimeters of mercury	Kgs/sq. meter	136.0

TO CONVERT	INTO	MULTIPLY BY
Centimeters of mercury	Lbs/sq. foot	27.85
Centimeters of mercury	Lbs/sq. inch	0.1934
Centimeters/second	Feet/min.	1.969
Centimeters/second	Feet/second	0.03281
Centimeters/second	Kilometers/hour	0.036
Centimeters/second	Meters/minute	0.6
Centimeters/second	Miles/hour	0.02237
Centimeters/second	Miles/minute	3.728×10^{-4}
Centimeters/sec/sec	Feet/sec/sec	0.03281
Centimeters/sec/sec	Kilometers/hr/sec	0.036
Centimeters/sec/sec	Miles/hr/sec	0.02237
Centimeters/sec/sec	Meters/sec/sec	0.01
Chains(surveyors)	Feet	66
Circular mils	Sq. centimeters	5.067×10^{-6}
Circular mils	Sq. inches	7.854×10^{-7}
Circular mils	Sq. mils	0.7854
Cord-feet	Cubic feet	4 ft $\times$ 4 ft $\times$ 1 ft
Cords	Cubic feet	8 ft $\times$ 4 ft $\times$ 4 ft
Coulombs	Abcoulombs	1/10
Coulombs	Statcoulombs	3×10^9
Coulombs	Faradays	1.036×10^{-5}
Coulombs/sq. inch	Abcoulombs/sq. cm.	0.01550
Coulombs/sq. inch	Coulombs/sq. cm.	0.1550
Coulombs/sq. inch	Statcoulombs/sq. cm.	4.65×10^8
Coulombs/sq. inch	Coulombs/sq. meter	1550
Coulombs/sq. cm.	Coulombs/sq. inch	64.52
Coulombs/sq. cm.	Coulombs/sq. meter	1×10^4
Coulombs/sq. meter	Coulombs/sq. cm.	1×10^{-4}
Coulombs/sq. meter	Coulombs/sq. inch	6.452×10^{-4}
Cubic Centimeters	Cubic feet	3.531×10^{-5}
Cubic centimeters	Cubic inches	6.102×10^{-2}
Cubic centimeters	Cubic meters	1×10^{-6}
Cubic centimeters	Cubic yards	1.308×10^{-6}
Cubic centimeters	Gallons (US liquid)	2.642×10^{-4}
Cubic centimeters	Liters	1×10^{-3}
Cubic centimeters	Pints (liquid)	2.113×10^{-3}
Cubic centimeters	Qts. (liquid)	1.057×10^{-3}
Cubic feet	Bushels (dry)	0.8036
Cubic feet	Cubic cm.	2.832×10^4
Cubic feet	Cubic in.	1728
Cubic feet	Cubic meters	0.02832
Cubic feet	Cubic yards	0.03704
Cubic feet	Gals. (liquid US)	7.481
Cubic feet	Liters	28.32
Cubic feet	Pts. (liquid)	59.84
Cubic feet	Qts. (liquid)	29.92
Cubic feet/minute	Cubic cm/sec.	472

TO CONVERT	INTO	MULTIPLY BY
Cubic feet/minute	Gallons/sec.	0.1247
Cubic feet/minute	Liters/sec.	0.4720
Cubic feet/minute	Lbs. of water/min.	62.4
Cubic feet/second	Millions gals/day	0.64632
Cubic feet/second	Gals/minute	448.8
Cubic inches	Cubic cm.	16.39
Cubic inches	Cubic ft.	5.787×10^{-4}
Cubic inches	Cubic meters	1.639×10^{-5}
Cubic inches	Cubic yards	2.143×10^{-5}
Cubic inches	Gals. (US liquid)	4.329×10^{-3}
Cubic inches	Liters	1.639×10^{-2}
Cubic inches	Mil-feet	1.061×10^{5}
Cubic inches	Pts. (liquid)	3.463×10^{-2}
Cubic inches	Qts. (liquid)	1.732×10^{-2}
Cubic meters	Bushels (dry)	28.38
Cubic meters	Cubic cm.	1×10^{6}
Cubic meters	Cubic feet	35.31
Cubic meters	Cubic inches	61,023
Cubic meters	Cubic yards	1.308
Cubic meters	Gallons	264.2
Cubic meters	Liters	1×10^{3}
Cubic meters	Pts. (liquid)	2,113
Cubic meters	Qts. (liquid)	1,057
Cubic yards	Cubic cm.	7.646×10^{5}
Cubic yards	Cubic feet	27
Cubic yards	Cubic inches	46,656
Cubic yards	Cubic meters	0.7646
Cubic yards	Gallons	202.0
Cubic yards	Liters	764.6
Cubic yards	Pts. (liquid)	1,616
Cubic yards	Qts. (liquid)	807.9
Cubic yards/minute	Cubic ft/sec	0.45
Cubic yards/minute	Gals/sec.	3.367
Cubic yards/minute	Liters/sec.	12.74
Cycles/second	Hertz	1.0
(D) Days	Hours	24
Days	Minutes	1,440
Days	Seconds	86,400
Decigrams	Grams	0.1
Deciliters	Liters	0.1
Decimeters	Meters	0.1
Degrees (angle)	Mils	17.45
Degrees (angle)	Minutes	60
Degrees (angle)	Quadrants	1.111×10^{-2}
Degrees (angle)	Radians	1.745×10^{-2}
Degrees (angle)	Seconds	3600

TO CONVERT	INTO	MULTIPLY BY
Degrees/second	Radians/sec	1.745×10^{-2}
Degrees/second	Revolutions/min.	0.1667
Degrees/second	Revolutions/sec.	2.778×10^{-3}
Dekagrams	Grams	10
Dekaliters	Liters	10
Dekameters	Meters	10
Drams	Grams	1.772
Drams	Grains	27.3437
Drams	Ounces	6.25×10^{-2}
Dynes	Grams	1.02×10^{-3}
Dynes	Joules/cm.	1×10^{-7}
Dynes	Poundals	7.233×10^{-5}
Dynes	Joules/meter	1×10^{-5}
Dynes	Pounds	2.248×10^{-6}
Dynes	Kilogram	1.02×10^{-6}
Dynes/square centimeter	Bars	1×10^{-6}
(E) Ergs	British Thermal units	9.486×10^{-11}
Ergs	dyne-centimeter	1.0
Ergs	Foot-pounds	7.376×10^{-8}
Ergs	Gram-calories	2.389×10^{-8}
Ergs	Gram-centimeter	1.02×10^{-3}
Ergs	Horsepower-hrs.	3.725×10^{-14}
Ergs	Joules	1×10^{-7}
Ergs	Kilogram-calories	2.39×10^{-11}
Ergs	Kilogram-meters	1.02×10^{-8}
Ergs	Kilowatt-hrs.	2.778×10^{-14}
Ergs	Watt-hours	2.778×10^{-11}
Ergs/second	B.T.U./min.	5.692×10^{-9}
Ergs/second	Ft-lbs/min.	4.426×10^{-6}
Ergs/second	Ft/lbs/sec.	7.376×10^{-8}
Ergs/second	Horsepower	1.341×10^{-10}
Ergs/second	Kilogram-cals/min	1.434×10^{-9}
Ergs/second	Kilowatts	1×10^{-10}
(F) Farads	Abfarads	1×10^{-9}
Farads	Microfarads	1×10^{6}
Farads	Picofarads	1×10^{12}
Farads	Statfarads	9×10^{11}
Faradays	Ampere-hours	26.8
Faradays	Coulombs	9.65×10^{4}
Fathoms	Feet	6.0
Feet	Centimeters	30.48
Feet	Inches	12.0
Feet	Kilometers	3.048×10^{-4}
Feet	Meters	3.048×10^{-1}
Feet	Miles (nautical)	1.645×10^{-4}

TO CONVERT	INTO	MULTIPLY BY
Feet	Miles (stat)	1.894×10^{-4}
Feet	Millimeters	3.048×10^{2}
Feet	Mils	1.2×10^{4}
Feet	Yards	0.333
Feet of water	Atmospheres	2.95×10^{-2}
Feet of water	Inches of mercury	0.8826
Feet of water	Kilograms/sq. cm.	3.048×10^{-2}
Feet of water	Kilograms/sq. meter	304.8
Feet of water	Lbs/sq. ft.	62.43
Feet of water	Lbs/sq. inch	4.335×10^{-1}
Feet/minute	Cm/sec	0.508
Feet/minute	Feet/sec	0.1667
Feet/minute	Kilometers/hr.	0.01829
Feet/minute	Meters/minute	0.3048
Feet/minute	Miles/hour	0.01136
Feet/second	Cm/second	30.48
Feet/second	Kilometers/hr	1.097
Feet/second	Meters/minute	18.29
Feet/second	Miles/hour	0.6818
Feet/second	Miles/minute	0.01136
Feet/100 feet	Percent grade	1.0
Feet/second per second	Cm/second/second	30.48
Feet/second per second	Kilometers/hr/sec	1.097
Feet/second per second	Meters/sec/sec	0.3048
Feet/second per second	Miles/hr/sec	0.6818
Foot-pounds	B.T.U.	1.286×10^{-3}
Foot-pounds	Ergs	1.356×10^{7}
Foot-pounds	Gram-calories	0.3238
Foot-pounds	Gram-cm	1.383×10^{4}
Foot-pounds	Horsepower hours	5.05×10^{-7}
Foot-pounds	Joules	1.356
Foot-pounds	Kilogram-calories	3.241×10^{-4}
Foot-pounds	Kilogram-meters	0.1383
Foot-pounds	Kilowatt-hours	3.766×10^{-7}
Foot-pounds	Ounce-inches	192
Foot-pounds/minute	B.T.U./minute	1.286×10^{-3}
Foot-pounds/minute	Ft-lbs/sec	0.01667
Foot-pounds/minute	Horsepower	3.03×10^{-5}
Foot-pounds/minute	Kilogram-cals/min	3.24×10^{-4}
Foot-pounds/minute	Kilowatts	2.26×10^{-5}
Foot-pounds/second	B.T.U./hour	4.6263
Foot-pounds/second	B.T.U./minute	0.07717
Foot-pounds/second	Horsepower	1.818×10^{-3}
Foot-pounds/second	Kilogram-cals/min	1.945×10^{-2}
Foot-pounds/second	Kilowatts	1.356×10^{-3}
(G) Gallons	Cubic centimeters	3785

TO CONVERT	INTO	MULTIPLY BY
Gallons	Cubic feet	0.1337
Gallons	Cubic inches	231
Gallons	Cubic meters	3.785×10^{-3}
Gallons	Cubic yards	4.951×10^{-3}
Gallons	Liters	3.785
Gallons	Pints (liquid)	8
Gallons	Quarts (liquid)	4
Gallons of water	Pounds of water	8.3453
Gallons/minute	Cubic ft/second	2.228×10^{-3}
Gallons/minute	Liters/second	0.06308
Gallons/minute	Cubic feet/hr	8.0208
Gausses	Lines/sq. inch	6.452
Gausses	Webers/sq. cm.	1×10^{-8}
Gausses	Webers/sq. inch	6.452×10^{-8}
Gausses	Webers/sq. meter	1×10^{-4}
Gilberts	Abampere-turns	0.07958
Gilberts	Ampere-turns	0.7958
Gilberts/centimeter	Ampere-turns/inch	2.021
Gilberts/centimeter	Ampere/turns/meter	79.58
Gills	Liters	0.1183
Gills	Pints (liquid)	0.25
Grains (troy)	Grains (avdp)	1.0
Grains (troy)	Grams	0.06480
Grains (troy)	Ounces (avdp)	2.0833×10^{-3}
Grains (troy)	Pennyweights (troy)	0.04167
Grains/U.S. gallon	Parts/million	17.118
Grains/U.S. gallon	Lbs/million gal	142.86
Grains/imperial gallon	Parts/million	14.286
Grams	Dynes	980.7
Grams	Grains (troy)	15.43
Grams	Joules/cm.	9.807×10^{-5}
Grams	Joules/meter	9.807×10^{-3}
Grams	Kilograms	1×10^{-3}
Grams	Milligrams	1×10^{3}
Grams	Ounces (avdp)	0.03527
Grams	Ounces (troy)	0.03215
Grams	Poundrals	0.07093
Grams	Pounds	2.205×10^{-3}
Gram-calories	British Thermal units	3.968×10^{-3}
Gram-calories	Ergs	4.1868×10^{7}
Gram-calories	Foot-pounds	3.0880
Gram-calories	Horsepower-hrs.	1.5596×10^{-6}
Gram-calories	Kilowatt-hours	1.1630×10^{-6}
Gram-calories	Watt-hours	1.1630×10^{-3}
Gram-calories/second	BTU/hour	14.286
Gram-centimeters	British Thermal Unit	9.302×10^{-8}
Gram-centimeters	Ergs	980.7

TO CONVERT	INTO	MULTIPLY BY
Gram-centimeters	Foot-pounds	7.233×10^{-5}
Gram-centimeters	Joules	9.807×10^{-5}
Gram-centimeters	Kilogram-calories	2.344×10^{-8}
Gram-centimeters	Kilogram-meter	1×10^{-5}
Grams/centimeter	Lbs/inch	5.6×10^{-3}
Grams/cubic centimeter	Lbs/cubic foot	62.43
Grams/cubic centimeter	Lbs/cubic inch	0.03613
Grams/cubic centimeter	Lbs/mil-foot	3.405×10^{-7}
Grams/liter	Grains/gallon	58.417
Grams/liter	Lbs/1000 gallon	8.345
Grams/liter	Lbs/cubic foot	6.243×10^{-2}
Grams/liter	Parts/million	1000
Grams/square centimeter	Lbs/sq. foot	2.0481
(H) Hectares	Acres	2.471
Hectares	Square feet	1.076×10^{5}
Hectograms	Grams	100
Hectoliters	Liters	100
Hectometers	Meters	100
Hectowatts	Watts	100
Hemispheres (solution angle)	Sphere	0.5
Hemispheres (solution angle)	Spherical right angles	4.0
Hemispheres (solution angle)	Steradians	6.283
Henries	Abhenries	1×10^{9}
Henries	Millihenries	1×10^{3}
Henries	Stathenries	$^{1}/_{9} \times 10^{-11}$
Hertz	Kilohertz	1×10^{-3}
Hertz	Megahertz	1×10^{-6}
Horsepower	BTU/minute	42.42
Horsepower	Ft/Lb/minute	33,000.0
Horsepower	Ft-lb/second	550.0
Horsepower	Horsepower (metric)	1.014
Horsepower	Kilogram-cal/min	10.7
Horsepower	Kilowatts	0.7457
Horsepower	Watts	745.7
Horsepower (metric)	Horsepower	0.9863
Horsepower (boiler)	BTU/hour	33,520
Horsepower (boiler)	Kilowatts	9.804
Horsepower-hours	BTU	2547
Horsepower-hours	Ergs	2.6845×10^{13}
Horsepower-hours	Foot-pounds	1.98×10^{6}
Horsepower-hours	Gram-calories	641,190.0
Horsepower-hours	Joules	2.684×10^{6}
Horsepower-hours	Kilogram-calories	641.7
Horsepower-hours	Kilogram-meters	2.737×10^{5}
Horsepower-hours	Kilowatt-hours	0.7457
Hours	Days	4.167×10^{-2}

	TO CONVERT	INTO	MULTIPLY BY
	Hours	Minutes	60.0
	Hours	Seconds	3600.0
	Hours	Weeks	5.952×10^{-3}
(I)	Inches	Centimeters	2.54
	Inches	Feet	8.333×10^{-2}
	Inches	Meters	2.54×10^{-2}
	Inches	Miles	1.578×10^{-5}
	Inches	Millimeters	25.40
	Inches	Mils	1000.0
	Inches	Yards	2.778×10^{-2}
	Inches of mercury	Atmospheres	0.03342
	Inches of mercury	Feet of water	1.133
	Inches of mercury	Kilograms/sq. meter	345.3
	Inches of mercury	Kilograms/sq. cm.	3.453×10^{-2}
	Inches of mercury	Lbs/sq. ft.	70.73
	Inches of mercury	Lbs/sq. inch	0.4912
	Inches of water (at 4° C)	Atmospheres	2.458×10^{-3}
	Inches of water (at 4° C)	Inches of mercury	7.355×10^{-2}
	Inches of water (at 4° C)	Kilograms/sq. cm.	2.54×10^{-3}
	Inches of water (at 4° C)	Kilograms/sq. meter	25.40
	Inches of water (at 4° C)	ounces/sq. inch	0.5781
	Inches of water (at 4° C)	Lbs/sq. foot	5.204
	Inches of water (at 4° C)	Lbs/sq. inch	0.03613
(J)	Joules	BTU	9.486×10^{-4}
	Joules	Ergs	1×10^{7}
	Joules	Foot-pounds	0.7376
	Joules	Kilogram-calories	2.389×10^{-4}
	Joules	Kilogram-meters	0.102
	Joules	Watt-hours	2.778×10^{-4}
	Joules/centimeter	Grams	1.02×10^{4}
	Joules/centimeter	Dynes	1×10^{7}
	Joules/centimeter	Joules/meter (Newtons)	100.0
	Joules/centimeter	Poundals	723.3
	Joules/centimeter	Pounds	22.48
(K)	Kilograms	Dynes	980,665.
	Kilograms	Grams	1000.0
	Kilograms	Joules/cm	0.09807
	Kilograms	Joules/meter (Newtons)	9.807
	Kilograms	Poundals	70.93
	Kilograms	Pounds (avdp)	2.2046
	Kilograms	Tons (long)	9.842×10^{-4}
	Kilograms	Tons (short)	1.102×10^{-3}
	Kilograms	Tonnes	1×10^{3}
	Kilogram-calories	BTU	3.968

TO CONVERT	INTO	MULTIPLY BY
Kilogram-calories	Foot-pounds	3086
Kilogram-calories	Horsepower-hours	1.558×10^{-3}
Kilogram-calories	Joules	4,186.0
Kilogram-calories	Kilogram-meters	426.6
Kilogram-calories	Kilojoules	4.186
Kilogram-calories	Kilowatt-hours	1.62×10^{-3}
Kilogram-calories/minute	Foot pounds/sec	51.43
Kilogram-calories/minute	Horsepower	0.09351
Kilogram-calories/minute	Kilowatts	0.06972
Kilogram-centimeters squared	Lbs-feet squared	2.373×10^{-3}
Kilogram-centimeters squared	Lbs-inches squared	0.3417
Kilogram-meters	BTU	9.302×10^{-3}
Kilogram-meters	Ergs	9.807×10^{7}
Kilogram-meters	Foot-pounds	7.233
Kilogram-meters	Joules	9.807
Kilogram-meters	Kilogram-calories	2.344×10^{-3}
Kilogram-meters	Kilowatt-hours	2.724×10^{-6}
Kilograms/cubic meter	Grams/cubic cm.	1×10^{-3}
Kilograms/cubic meter	Lbs/cubic foot	0.06243
Kilograms/cubic meter	Lbs/cubic inch	3.613×10^{-5}
Kilograms/cubic meter	Lbs/mili-foot	3.405×10^{-10}
Kilograms/meter	Lbs/foot	0.6720
Kilograms/sq. meter	Atmospheres	9.678×10^{-5}
Kilograms/sq. meter	Bars	98.07×10^{-6}
Kilograms/sq. meter	Feet of water	3.281×10^{-3}
Kilograms/sq. meter	Inches of mercury	2.896×10^{-3}
Kilograms/sq. meter	Lbs/sq. foot	0.2048
Kilograms/sq. meter	Lbs/sq. inch	1.422×10^{-3}
Kilograms/sq. millimeter	Kilograms/sq. meter	1×10^{6}
Kilolines	Maxwells	1×10^{3}
Kiloliters	Liters	1×10^{3}
Kilometers	Centimeters	1×10^{5}
Kilometers	Feet	3281
Kilometers	Inches	3.937×10^{4}
Kilometers	Light years	1.0567×10^{-13}
Kilometers	Meters	1×10^{3}
Kilometers	Miles	0.6214
Kilometers	Millimeters	1×10^{6}
Kilometers	Yards	1093.6
Kilometers/hour	Cm/second	27.78
Kilometers/hour	Feet/minute	54.68
Kilometers/hour	Feet/second	0.9113
Kilometers/hour	Meters/min.	16.67
Kilometers/hour	Miles/hour	0.6214
Kilometers/hour/second	Cm/sec/sec	27.78
Kilometers/hour/second	Feet/sec/sec	0.9113
Kilometers/hour/second	Meters/sec/sec	0.2778

TO CONVERT	INTO	MULTIPLY BY
Kilometers/hour/second	Miles/hr/sec	0.6214
Kilometers/minute	Kilometers/hr.	60.0
Kilowatts	BTU/minute	56.92
Kilowatts	Foot-lbs/minute	4.425×10^4
Kilowatts	Foot-lbs/second	737.6
Kilowatts	Horsepower	1.341
Kilowatts	Kilogram-cals/min	14.34
Kilowatts	Watts	1×10^3
Kilowatts-hours	BTU	3415
Kilowatt-hours	Ergs	3.6×10^{13}
Kilowatt-hours	Foot-pounds	2.655×10^6
Kilowatt-hours	Gram-calories	859,850.
Kilowatt-hours	Horsepower-hours	1.341
Kilowatt-hours	Joules	3.6×10^6
Kilowatt-hours	Kilogram-cals	860.5
Kilowatt-hours	Kilogram-meters	3.671×10^5
Kilowatt-hours	Watt-hours	1×10^3
(L) Lamberts	Candles/sq. cm.	0.3183
Lamberts	Candles/sq. inch	2.054
League	Miles (approx.)	3.0
Lines/sq. centimeter	Gausses	1.0
Lines/sq. inch	Gausses	0.1550
Lines/sq. inch	Webers/sq. cm.	1.55×10^{-9}
Lines/sq. inch	Webers/sq. inch	1×10^{-8}
Lines/sq. inch	Webers/sq. meter	1.55×10^{-5}
Links	Chains	0.01
Links (Engineers)	Inches	12.0
Links (Surveyors)	Inches	7.92
Liters	Bushels (US dry)	0.0284
Liters	Cubic cm.	1×10^3
Liters	Cubic feet	0.03531
Liters	Cubic inches	61.02
Liters	Cubic meters	1×10^{-3}
Liters	Cubic yards	1.308×10^{-3}
Liters	Gals (US liquid)	0.2642
Liters	Pts. (US liquid)	2.113
Liters	Quarts (US liquid)	1.057
Liters/minute	Cubic feet/sec	5.885×10^{-4}
Liters/minute	Gals/second	4.403×10^{-3}
Lumens/sq. foot	Foot-candles	1.0
Lux	Foot-candles	0.0929
(M) Maxwells	Kilolines	1×10^{-3}
Maxwells	Megalines	1×10^{-6}
Maxwells	Webers	1×10^{-8}
Megalines	Maxwells	1×10^6

TO CONVERT	INTO	MULTIPLY BY
Megmhos/centimeter cube	Abmhos/cm. cube	1×10^{-3}
Megmhos/centimeter cube	Megmhos/inch cube	2.54
Mehmhos/centimeter cube	Mhos/mil foot	0.1662
Megmhos/inch cube	Megmhos/cm. cube	0.3937
Megohms	Ohms	1×10^{6}
Megohms	Microhms	1×10^{12}
Meters	Centimeters	100
Meters	Feet	3.2808
Meters	Inches	39.37
Meters	Kilometers	1×10^{-3}
Meters	Millimeters	1×10^{3}
Meters	Miles (nautical)	5.396×10^{-4}
Meters	Miles (statute)	6.214×10^{-4}
Meters	Yards	1.094
Meter-Kilograms	Centimeter-dynes	9.807×10^{7}
Meter-Kilograms	Centimeter-grams	1×10^{5}
Meter-Kilograms	Pound-feet	7.233
Meters/minute	Cm/sec	1.667
Meters/minute	Ft/minute	3.281
Meters/minute	Ft/second	0.05468
Meters/minute	Kilometers/hr	0.06
Meters/minute	Miles/hr	0.03728
Meters/second	Ft/minute	196.8
Meters/second	Ft/second	3.281
Meters/second	Kilometers/hr	3.6
Meters/second	Kilometers/min	0.06
Meters/second	Miles/hour	2.237
Meters/second	Miles/minute	0.03728
Meters/sec/sec	Cm/sec/sec	100.0
Meters/sec/sec	Ft/sec/sec	3.281
Meters/sec/sec	Kilometers/hr/sec	3.6
Meters/sec/sec	Miles/hr/sec	2.237
Mhos	Micromhos	1×10^{6}
Mhos	Millimhos	1×10^{3}
Microfarads	Abfarads	1×10^{-15}
Microfarads	Farads	1×10^{-6}
Microfarads	Picofarads	1×10^{6}
Microfarads	Statfarads	9×10^{5}
Micrograms	Grams	1×10^{-6}
Microhms	Abohms	1×10^{3}
Microhms	Megohms	1×10^{-12}
Microhms	Ohms	1×10^{-6}
Microhms	Statohms	$^{1}/_{9} \times 10^{-17}$
Microhms/centimeter cube	Abohms/cm cube	1×10^{3}
Microhms/centimeter cube	Microhms/inch cube	0.3937
Microhms/centimeter cube	Ohms/mil foot	6.015
Microhms/inch cube	Microhms/cm cube	2.540

TO CONVERT	INTO	MULTIPLY BY
Microliters	Liters	1×10^{-6}
Microns	Meters	1×10^{-6}
Miles (nautical)	Feet	6,076.103
Miles (nautical)	Kilometers	1.852
Miles (nautical)	Meters	1,852.
Miles (nautical)	Miles (statute)	1.1508
Miles (nautical)	Yards	2,025.4
Miles (statute)	Centimeters	1.609×10^{5}
Miles (statute)	Feet	5,280
Miles (statute)	Inches	6.336×10^{4}
Miles (statute)	Kilometers	1.609
Miles (statute)	Light years	1.691×10^{-13}
Miles (statute)	Meters	1,609.0
Miles (statute)	Miles (nautical)	0.8684
Miles (statute)	Yards	1760.0
Miles/hour	Centimeters/sec	44.7
Miles/hour	Feet/minute	88
Miles/hour	Feet/second	1.467
Miles/hour	Kilometers/hour	1.6093
Miles/hour	Kilometers/minute	0.0268
Miles/hour	Meters/minute	26.82
Miles/hour	Miles/minute	0.01667
Miles/hour/second	Centimeters/sec/sec	44.7
Miles/hr/hr/sec	Feet/sec/sec	1.467
Miles/hr/hr/sec	Kilometers/hr/sec	1.6093
Miles/hr/hr/sec	Meters/sec/sec	0.447
Miles/minute	Centimeters/sec	2682
Miles/minute	Feet/second	88
Miles/minute	Kilometers/min	1.6093
Miles/minute	Miles/hour	60
Milliamperes	Microamperes	1×10^{3}
Milliers	Kilograms	1×10^{3}
Milligrams	Grams	1×10^{-3}
Milligrams/liter	Parts/million	1.0
Millihenries	Abhenries	1×10^{6}
Millihenries	Henries	1×10^{-3}
Millihenries	Microhenries	1×10^{3}
Millihenries	Stathenries	$^{1}/_{9} \times 10^{-14}$
Milliliters	Liters	1×10^{-3}
Millimeters	Centimeters	0.1
Millimeters	Feet	3.281×10^{-3}
Millimeters	Inches	0.03937
Millimeters	Kilometers	1×10^{-6}
Millimeters	Meters	1×10^{-3}
Millimeters	Microns	1×10^{3}
Millimeters	Miles	6.214×10^{-7}
Millimeters	Mils	39.37

TO CONVERT	INTO	MULTIPLY BY
Millimeters	Yards	1.094×10^{-3}
Million gallons/day	Cubic ft/sec	1.54723
Millivolts	Microvolts	1×10^{3}
Mils	Centimeters	2.54×10^{-3}
Mils	Feet	8.333×10^{-5}
Mils	Inches	1×10^{-3}
Mils	Kilometers	2.54×10^{-8}
Mils	Minutes	3.438
Mils	Yards	2.778×10^{-5}
Miner's inches	Cubic feet/min	1.5
Minutes (angle)	Degrees	0.01667
Minutes (angle)	Quadrants	1.852×10^{-4}
Minutes (angle)	Radians	2.909×10^{-4}
Minutes (angle)	Seconds	60
Months	Days	30.42
Months	Hours	730
Months	Minutes	43,800
Months	Seconds	2.628×10^{6}
Myriagrams	Kilograms	10
Myriameters	Kilometers	10
Myriawatts	Kilowatts	10
(N) Nepers	Decibels	8.686
Newtons	Dynes	1×10^{5}
Newtons	Pounds (avdp)	0.2248
(O) Ohms	Abohms	1×10^{9}
Ohms	Ohms (international)	0.99948
Ohms	Megohms	1×10^{-6}
Ohms	Microhms	1×10^{6}
Ohms	Milliohms	1×10^{3}
Ohms	Statohms	$^{1}/_{9} \times 10^{-11}$
Ohms/foot	Ohms/meter	0.03048
Ohms/mil foot	Abohms/cm. cube	166.2
Ohms/mil foot	Microhms/cm. cube	0.1662
Ohms/mil foot	Microhms/inch cube	0.06524
Ounces	Drams	8
Ounces	Grains	437.5
Ounces	Grams	28.35
Ounces	Pounds	0.0625
Ounces	Ounces (troy)	0.9115
Ounces	Ton (long)	2.790×10^{-5}
Ounces	Ton (metric)	2.835×10^{-5}
Ounces (fluid)	Cubic inches	1.805
Ounces (fluid)	Liters	0.02957
Ounces (fluid)	Quarts	0.03125
Ounces (troy)	Grains (troy)	480

	TO CONVERT	INTO	MULTIPLY BY
	Ounces (troy)	Grams	31.10
	Ounces (troy)	Ounces (avdp)	1.09714
	Ounces (troy)	Pennyweights (troy)	20.0
	Ounces (troy)	Pounds (troy)	0.08333
	Ounces/square inch	Lbs/sq. inch	0.0625
(P)	Parts/million	Grains/U.S. gallon	0.0584
	Parts/million	Grains/imperial gal.	0.07016
	Parts/million	Lbs/million gallon	8.345
	Pennyweights (troy)	Grains (troy)	24
	Pennyweights (troy)	Grams (troy)	1.555
	Pennyweights (troy)	Ounces (troy)	0.05
	Pennyweights (troy)	Pounds (troy)	4.1667×10^{-3}
	Perches (masonry)	Cubic feet	24.75
	Picofarad	Micromicrofarad	1.0
	Pints (dry)	Cubic inches	33.6
	Pints (liquid)	Cubic cm.	473.2
	Pints (liquid)	Cubic feet	0.0167
	Pints (liquid)	Cubic inches	28.87
	Pints (liquid)	Cubic meters	4.732×10^{-4}
	Pints (liquid)	Cubic yards	6.189×10^{-4}
	Pints (liquid)	Gallons	0.125
	Pints (liquid)	Liters	0.4732
	Pints (liquid)	Quarts (liquid)	0.5
	Poundals	Dynes	13,826
	Poundals	Grams	14.1
	Poundals	Joules/cm.	1.383×10^{-3}
	Poundals	Joules/meter	0.1383
	Poundals	Kilograms	0.0141
	Poundals	Pounds (avdp)	0.03108
	Pounds	Drams	256
	Pounds	Dynes	44.48×10^{4}
	Pounds	Grains	7000
	Pounds	Grams	453.59
	Pounds	Joules/cm	0.04448
	Pounds	Joules/meter	4.448
	Pounds	Kilograms	0.4536
	Pounds	Ounces	16.0
	Pounds	Ounces (troy)	14.5833
	Pounds	Poundals	32.17
	Pounds	Pounds (troy)	1.2153
	Pounds	Tons (short)	5×10^{-4}
	Pounds (troy)	Grains	5760
	Pounds (troy)	Grams	373.242
	Pounds (troy)	Ounces (avdp)	13.157
	Pounds (troy)	Ounces (troy)	12.0
	Pounds (troy)	Pennyweight (troy)	240

TO CONVERT	INTO	MULTIPLY BY
Pounds (troy)	Pounds (avdp)	0.82286
Pounds (troy)	Tons (long)	3.6735×10^{-4}
Pounds (troy)	Tons (metric)	3.7324×10^{-4}
Pounds (troy)	Tons (short)	4.1143×10^{-4}
Pounds of water	Cubic feet	0.01602
Pounds of water	Cubic inches	27.68
Pounds of water	Gallons	0.1198
Pounds of water/minute	Cubic ft/sec	2.67×10^{-4}
Pound-feet	Cm-dynes	1.356×10^7
Pound-feet	Cm-grams	13,825
Pound-feet	Meter-kilograms	0.1383
Pounds-feet squared	Kilogram-cm. squared	421.3
Pounds-feet squared	Lbs-inches squared	144
Pounds-inches squared	Kilograms-cm squared	2.926
Pounds-inches squared	Pounds-feet squared	6.945×10^{-3}
Pounds/cubic foot	Grams/cubic cm.	0.01602
Pounds/cubic foot	Kilograms/cubic meter	16.02
Pounds per cubic foot	Lbs/cubic inch	5.787×10^{-4}
Pounds/cubic foot	Lbs/mil foot	5.456×10^{-9}
Pounds/cubic inch	Grams/cubic cm	27.68
Pounds/cubic inch	Kilograms/cubic meter	2.768×10^4
Pounds/cubic inch	Pounds/cubic foot	1728
Pounds/cubic inch	Pounds/mil foot	9.425×10^{-6}
Pounds/foot	Kilograms/meter	1.488
Pounds/inch	Grams/cm.	178.6
Pounds/mil foot	Grams/cubic cm.	2.306×10^6
Pounds/square foot	Atmospheres	4.725×10^{-4}
Pounds/square foot	Feet of water	0.01602
Pounds/square foot	Inches of mercury	0.01414
Pounds/square foot	Kilograms/sq. meter	4.882
Pounds/square foot	Pounds/sq. inch	6.944×10^{-3}
Pounds/square inch	Atmospheres	0.06804
Pounds/square inch	Feet of water	2.307
Pounds/square inch	Inches of mercury	2.036
Pounds/square inch	Kilograms/sq. meter	703.1
Pounds/square inch	Lbs/sq. foot	144
(Q) Quadrants (angle)	Degrees	90
Quadrants (angle)	Minutes	5400
Quadrants (angle)	Radians	1.571
Quadrants (angle)	Seconds	3.24×10^5
Quarts (dry)	Cubic inches	67.2
Quarts (liquid)	Cubic cm.	946.4
Quarts (liquid)	Cubic feet	0.03342
Quarts (liquid)	Cubic inches	57.75
Quarts (liquid)	Cubic meters	9.464×10^{-4}
Quarts (liquid)	Cubic yards	1.238×10^{-3}

TO CONVERT	INTO	MULTIPLY BY
Quarts (liquid)	Gallons	0.25
Quarts (liquid)	Liters	0.9463
Quintals	Pounds	100
Quires	Sheets	25
(R) Radians	Degrees	57.3
Radians	Mils	1×10^3
Radians	Minutes	3438
Radians	Quadrants	0.637
Radians	Seconds	2.063×10^5
Radians/second	Degrees/second	57.3
Radians/second	Revolutions/min	9.549
Radians/second	Revolutions/sec	0.1592
Radians/sec/sec	Rev/min/min	573.0
Radians/sec/sec	Rev/min/sec	9.549
Radians/sec/sec	Rev/sec/sec	0.1592
Revolutions	Degrees	360
Revolutions	Quadrants	4
Revolutions	Radians	6.283
Revolutions/minute	Degrees/sec	6
Revolutions/minute	Radians/second	0.1047
Revolutions/minute	Rev/second	0.01667
Revolutions/min/min	Radians/sec/sec	1.745×10^{-3}
Revolutions/min/min	Rev/min/sec	0.01667
Revolutions/min/min	Rev/sec/sec	2.778×10^{-4}
Revolutions/second	Degrees/second	360
Revolutions/second	Radians/second	6.283
Revolutions/second	Rev/minute	60
Revolutions/sec/sec	Radians/sec/sec	6.283
Revolutions/sec/sec	Rev/min/min	3600
Revolutions/sec/sec	Rev/min/sec	60
Rods	Feet	16.5
Rods	Miles	3.125×10^{-3}
Rods	Yards	5.5
(S) Seconds (angle)	Degrees	2.778×10^{-4}
Seconds (angle)	Minutes	0.01667
Seconds (angle)	Quadrants	3.087×10^{-6}
Seconds (angle)	Radians	4.848×10^{-6}
Spheres (solid angle)	Steradians	12.57
Spherical right angles	Hemispheres	0.25
Spherical right angles	Spheres	0.125
Spherical right angles	Steradians	1.571
Square centimeters	Circular mils	1.973×10^5
Square centimeters	Square feet	1.076×10^{-3}
Square centimeters	Square inches	0.155
Square centimeters	Square meters	1×10^{-6}
Square centimeters	Square miles	3.861×10^{-11}

TO CONVERT	INTO	MULTIPLY BY
Square centimeters	Square millimeters	100
Square centimeters	Square yards	1.196×10^{-4}
Square feet	Acres	2.296×10^{-5}
Square feet	Circular mils	1.833×10^{8}
Square feet	Square cm.	929
Square feet	Square inches	144
Square feet	Square meters	0.0929
Square feet	Square miles	3.587×10^{-8}
Square feet	Sq. millimeters	9.29×10^{4}
Square feet	Square yards	0.1111
Square ft-ft squared	Sq. inches-inches sq.	2.074×10^{4}
Square inches	Circular mils	1.273×10^{6}
Square inches	Square cm.	6.452
Square inches	Square feet	6.944×10^{-3}
Square inches	Square mils	1×10^{6}
Square inches	Square millimeters	645.2
Square inches	Square yards	7.716×10^{-4}
Square inches-inches squared	Sq. cm-cm squared	41.62
Square inches-inches squared	Sq ft-ft. squared	4.823×10^{-5}
Square kilometers	Acres	247.1
Square kilometers	Sq. centimeters	1×10^{10}
Square kilometers	Square feet	10.76×10^{6}
Square kilometers	Square inches	1.55×10^{9}
Square kilometers	Square meters	1×10^{6}
Square kilometers	Square miles	0.3861
Square kilometers	Square yards	1.196×10^{6}
Square meters	Acres	2.471×10^{-4}
Square meters	Sq. centimeters	1×10^{4}
Square meters	Square feet	10.764
Square meters	Sq. inches	1555
Square meters	Sq. miles	3.861×10^{-7}
Square meters	Sq. millimeters	1×10^{6}
Square meters	Sq. yards	1.196
Square miles	Acres	640
Square miles	Sq. feet	27.88×10^{6}
Square miles	Sq. kilometers	2.59
Square miles	Sq. meters	2.59×10^{6}
Square miles	Sq. yards	3.098×10^{6}
Square millimeters	Circular mils	1.973×10^{3}
Square millimeters	Sq. centimeters	0.01
Square millimeters	Sq. feet	1.076×10^{-5}
Square millimeters	Sq. inches	1.55×10^{-3}
Square mils	Circular mils	1.273
Square mils	Sq. cm.	6.452×10^{-6}
Square mils	Sq. inches	1×10^{-6}
Square yards	Acres	2.066×10^{-4}
Square yards	Sq. cm.	8361
Square yards	Sq. feet	9

TO CONVERT	INTO	MULTIPLY BY
Square yards	Sq. inches	1296
Square yards	Sq. meters	0.8361
Square yards	Sq. miles	3.228×10^{-7}
Square yards	Sq. millimeters	8.361×10^5
Statamperes	Abamperes	$1/3 \times 10^{-10}$
Statamperes	Amperes	$1/3 \times 10^{-9}$
Statcoulombs	Abcoulombs	$1/3 \times 10^{-10}$
Statcoulombs	Coulombs	$1/3 \times 10^{-9}$
Statfarads	Abfarads	$1/9 \times 10^{-20}$
Statfarads	Farads	$1/9 \times 10^{-11}$
Statfarads	Microfarads	$1/9 \times 10^{-5}$
Stathenries	Abhenries	9×10^{20}
Stathenries	Henries	9×10^{11}
Stathenries	Millihenries	9×10^{14}
Statohms	Abohms	9×10^{20}
Statohms	Megohms	9×10^5
Statohms	Microhms	9×10^{17}
Statohms	Ohms	9×10^{11}
Statvolts	Abvolts	3×10^{10}
Statvolts	Volts	300
Steradians	Hemispheres	0.1592
Steradians	Spheres	0.07958
Steradians	Spherical right angles	0.6366
Steres	Liters	1×10^3
(T) Tons (long)	Kilograms	1016
Tons (long)	Pounds	2240
Tons (long)	Tons (short)	1.12
Tons (metric)	Kilograms	1000
Tons (metric)	Pounds	2205
Tons (short)	Kilograms	907.2
Tons (short)	Ounces	32000
Tons (short)	Ounces (troy)	29,166.66
Tons (short)	Pounds	2000
Tons (short)	Pounds (troy)	2430.6
Tons (short)	Tons (long)	0.8928
Tons (short)	Tons (metric)	0.9078
Tons (short)/square foot	Kilograms/sq. meter	9765
Tons (short)/square foot	Lbs/square inch	13.89
Tons (short)/sq. inch	Kilograms/sq. meter	1.406×10^6
Tons (short)/square inch	Lbs/sq. inch	2000
Tons of water/24 hrs.	Lbs of water/hr.	83.333
Tons of water/24 hours	Gallons/min	0.16643
Tons of water/24 hours	Cubic feet/hr	1.3349
(V) Varas	Feet	2.7777
Varas	Inches	33.333

TO CONVERT	INTO	MULTIPLY BY
Varas	Miles	5.26×10^{-4}
Varas	Yards	0.9259
Volts	Abvolts	1×10^8
Volts	Kilovolts	1×10^{-3}
Volts	Microvolts	1×10^6
Volts	Millivolts	1×10^3
Volts	Statvolts	1/300
Volts/inch	Abvolts/cm.	3.937×10^7
Volts/inch	Statvolts/cm.	1.312×10^{-3}
(W) Watts	BTU/min.	0.05692
Watts	BTU/hr.	3.413
Watts	Ergs/second	1×10^7
Watts	Ft-lb/minute	44.26
Watts	Ft-lbs/second	0.7376
Watts	Horsepower	1.341×10^{-3}
Watts	Horsepower (metric)	1.360×10^{-3}
Watts	Kilogram-cals/min	0.01434
Watts	Kilowatts	1×10^{-3}
Watts	Microwatts	1×10^6
Watts	Milliwatts	1×10^3
Watt-hours	BTU	3.415
Watt-hours	Ergs	3.6×10^{10}
Watt-hours	Foot-pounds	2655
Watt-hours	Gram-cals	859.85
Watt-hours	Horsepower-hrs	1.341×10^{-3}
Watt-hours	Kilogram-cals	0.8605
Watt-hours	Kilogram-meters	367.2
Watt-hours	Kilowatt-hrs	1×10^{-3}
Webers	Maxwells	1×10^8
Webers	Kilolines	1×10^5
Webers/square inch	Gausses	1.55×10^7
Webers/square inch	Lines/sq. inch	1×10^8
Webers/square inch	Webers/sq. cm.	0.155
Webers/square inch	Webers/sq. meter	1550
Webers/square meter	Gausses	1×10^4
Webers/square meter	Lines/sq. inch	6.452×10^4
Webers/square meter	Webers/sq. cm.	1×10^{-4}
Webers/square meter	Webers/sq. inch	6.452×10^{-4}
Weeks	Hours	168
Weeks	Minutes	10,080.
Weeks	Seconds	604,800.
(Y) Yards	Centimeters	91.44
Yards	Feet	3.0
Yards	Inches	36.0
Yards	Kilometers	9.144×10^{-4}

TO CONVERT	INTO	MULTIPLY BY
Yards	Meters	0.9144
Yards	Miles (nautical)	4.934×10^{-4}
Yards	Miles (statute)	5.682×10^{-4}
Yards	Millimeters	914.4
Years (common)	Days	365
Years (common)	Hours	8760
Years (leap)	Days	366
Years (leap)	Hours	8784

CHAPTER 7

Troubleshooting Through
Circuitry Design Analysis

If one were to ask what is Electronics really all about you might answer that it consists of three circuits: series, parallel, and series-parallel and five groups of components: resistors, capacitors, inductors, vacuum tubes and solid state devices. It is the arrangement of these components with their inherent characteristics and the three types of circuitry which produces the desired electronic result. Hence, the difference between a radio and TV or between any other electronic device is their circuitry-component arrangement.

In this chapter a common type, transformerless superheterodyne radio receiver will be dissected and partially reconstructed in a down-to-earth, step-by-step approach to circuitry-component-purpose arrangement. The stages we will closely examine will be the power supply, input, and output. For the purpose of clarifying the objective, the desired electronic outcome will be a 1-watt, 3-tube AC/DC amplifier.

All electronic devices can be broken down into smaller segments or blocks representing stages or circuitry divisions. In our amplifier's case, only 3 divisions will be necessary. These three, as illustrated in Figure 7-1, include the input stage, the output stage, and the power supply.

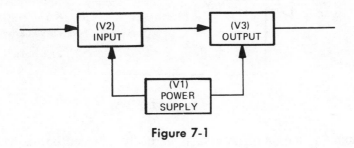

Figure 7-1

7.1 POWER SUPPLY DESIGN

The manufacturers of electronic devices are cost minimizers, and therefore will avoid costly inductors and transformers in circuitry in which a power supply of a lessor cost could be employed. This is done, naturally, if it does not lower the quality of their particular device. The full-wave, bridge type, transformer-powered circuitry is replaced with two diodes thereby keeping full-wave rectification and the transformer.

Often these two diodes may be replaced with a single diode. The latter is more economical because only one diode is employed and no transformer is required. This type is called an AC/DC power supply and is the one we are specifically interested in.

Chart 7-1 will provide you with the particulars we will be requiring throughout some of the power supply design. The tube used for the output amplifier stage is a 50C5 called a beam power tube. The tube in the input stage is a 12AV6, called a twin diode, high-mu (gain) triode. If this were a radio, the two diode plates would be used as the detector, but since we are only designing the amplifier portion, the detector-diode plates will not be considered. The third and perhaps most important tube is the 35W4, a half-wave rectifier.

TUBE	TYPICAL B +	HEATER VOLTAGE	HEATER CURRENT
35W4 (V1)	360 VDC PIV	35 VDC/VAC	0.15 amperes
12AV6 (V2)	250 VDC	12.6 VDC/VAC	0.15
50C5 (V3)	120 VDC	50 VDC/VAC	0.15 amperes
	150 VDC MAX		

Chart 7-1

The one item not shown in Chart 7-1 is the normal AC plate supply voltage for the 35W4 of 117 VAC. This is important because it is this potential that we wish to rectify.

Electronic manufactors, when using an AC/DC power supply must be careful to select electronic devices whose current through them is the same value to eliminate possible destruction. The series circuit illustrated in Figure 7-2 or similar is normally used. All three tubes we are using require 150 mA of current as noted in Chart 7-1, to heat up the filaments.

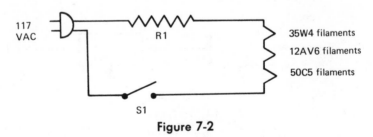

Figure 7-2

Switch S1, the on-off switch, is normally mounted on the back side of the volume control. Since Figure 7-2 illustrates a series circuit, then we already know that the total current flowing in the filament circuit after the switch is closed is 150 mA. We now must determine how much of the applied 117 volts AC is used by the vacuum tubes filaments. This is accomplished simply by adding their filament voltages.

$$
\begin{array}{ll}
\text{(V1) 35W4} & \text{.................35.0 volts} \\
\text{(V2) 12AV6} & \text{.................12.6 volts} \\
\text{(V3) 50C5} & \text{.................\underline{50.0 volts}} \\
& \text{97.6 volts}
\end{array}
$$

As our calculations indicate, we do not use 19.4 volts. This naturally is the voltage to be dropped across resistor R_1 of Figure 7-2. Resistor R_1, called a current limiting resistor, must absorb this unused potential or else we would have to replace the tubes quite frequently because they would burn up. The resistive value for R_1 is a snap to calculate since we know the series filament circuit has 150mA flowing, the value for I, and the voltage, the value for E, dropped across it should be 19.4 volts, then

$$R_1 = \frac{ER_1}{IR_1} = \frac{19.4v}{150mA} = 129.33 \text{ ohms}$$

Now, if you think the manufacturer will run out and buy x amounts of our design value resistors, you are quite incorrect. He will buy inexpensive resistors which are mass produced and available in preferred values as set up by the Electronics Industries Association (EIA), formally RETMA. If you were to look up what value came closest to our design value in the 10-20% tolerance power type resistor data, the manufacturer's choice would be 133 ohms or 140 ohms or 150 ohms value, depending upon which group he could procure cheapest. Normally the 20% tolerance is cheapest. We shall use 150 ohms, first for ease in calculations and second, as a rule of thumb, the next larger EIA value is selected, and third, an educated guess as to probable realistic choice.

The power rating also must be attached to the 150 ohm resistor before ordering. This, however, could prove confusing if you compared the three results from the three power determining formulas: $P = IE$, $P = I^2R$, and $P = E^2/R$. This is evident since the designed value of 129 ohms was changed to 150 ohms, and in doing so the 150 mA current value would decrease because the only constant would ahve been the 19.4 volts. Note chart 7-2 for this comparison.

$P = IE$	$P = I^2R$	$P = \dfrac{E^2}{R}$
2.91 Watt	3.375 watt	2.509 watt

Chart 7-2

From the data contained in the power calculation comparison we can determine that the minimum wattage would be 2.5 watts to a maximum of 3.4 watts (rounded off). Each resistor's wattage rating is evaluated assuming standard ambient temperature facts, a specific life expectancy, and a stated long-term drift from a no-load value. The chances are good that a 4-watt resistor would work, but a 5-watt would be better and perhaps easier to come by. We, therefore, shall make the value of R_1 equal to 150 ohms at 5 watts.

All of this was necessary just to allow the filament circuit to function as desired. We now shall design the rest of the power supply circuitry so that B + can be obtained. The rest of the power supply and related important circuitry for design calculations are drawn in Figure 7-3 and will serve for reference in the following guidelines.

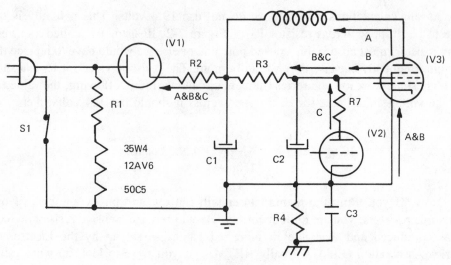

Figure 7-3

The three illustrated current paths labeled A, B and C shall be used to determine the values for resistors R_2 and R_3 and eventually R_7. Resistor R_2, called a current limiting resistor, has the main purpose of protecting the halfwave rectifier from self-destruction. Before this resistor is determined we should make certain of the total current capabilities of all three tubes. Chart 7-3 shows tube manual specifics for current limitations.

(V1) 35W4	(V2) 12AV6	(V3) 50C5
100mA average output current	1.2 mA max plate current	50 mA max plate current 8 mA max grid #2 current

Chart 7-3

The designer should make certain that the total current flowing through or to the R_2-R_3 connection does not exceed the current limitations for V_1. In our case we are well under this since currents A = 50mA, plus B = 8mA, plus C = 1.2mA equals only 59.2 mA a value well under tube manual limitations for the 35W4.

It is desirable, therefore, to have the value of R_2 to be such that it WILL burn out before the maximum average current of 100 mA flows through V_1. Of course, manufacturers won't let the resistor's wattage value be too low so that it will burn out under normal operating conditions, therefore a one-quarter watt is out. Normally a one-half watt resistor will provide protection for the rectifier and also the fuse action should it be needed. To determine its value we simply plug in the known or desired specifics. Wattage equals 0.5 and V_1's total current equals 100mA so . . .

$$R_2 = \frac{P}{I^2} = \frac{0.5}{(0.1)^2} = 50 \text{ ohms}$$

Again, as a general rule of thumb for this particular resistor, a value of the next larger EIA (RETMA) value is selected. This value is 56 ohms at one-half watt, if easily obtainable, or (68) ohms if not. This selection was from a 10% tolerance resistor grouping, but a 68-ohm resistor having a 20% tolerance could be used thereby keeping the expense down.

7.1-1   Filtering Circuitry

The filter circuitry consists of the input filter capacitor C_1, the filter resistor R_3, and the output filter capacitor C_2. This type of filter is called a "Pi" filter because its arrangement of C_1, R_3 and C_2 resembles the Greek symbol π. Most power supply filter capacitors are of the electrolytic variety because better filtering is obtained when using large valued capacitors. The larger the value, the better filtering, but also the more costly. The trusty tube manual recommends an input capacitor of 40mF, but its voltage rating is left, I guess, to your imagination. Working voltage for electrolytic capacitors are quite important unless you like having fireworks on your bench. This is what it is all about: the maximum voltage across the input capacitor C_1 is the peak value or maximum value of the AC line voltage. The maximum voltage is calculated as shown.

$$E \text{ max} = 1.414 \times 117V = 165.438 \text{ volts}$$

It is apparent from Figure 7-4 that the 117 VAC line voltage is applied to or across the series circuit made up of V_1, R_2, and C_1. The voltage across the input filter capacitor will be somewhat lower than the maximum voltage just calculated, in fact, a typical voltage would be the RMS value of the 117 volt line. A quite reasonable voltage rating for C_1 would be 150 VDC because coincidentally this is the maximum plate voltage rating given for the 12AV6 and it is high enough in value when compared to the rms value of the line voltage.

The manufacturer would probably buy one filter capacitor having two units of capacitance rated at 150 volts DC. The basis for doing this perhaps is a reduction of inventory and a chance to juggle their specific capacitance values depending upon availability. Common filter capacitors would be 40-40 at 150VDC, 50-30 at 150 VDC,

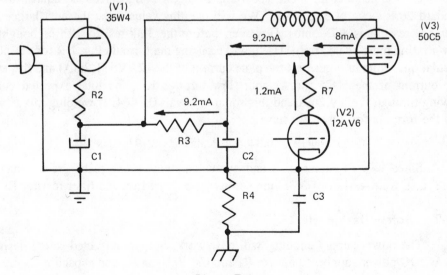

Figure 7-4

or 50-50 at 150 VDC. Most tube manuals recommend the input filter to be 40 μ F. Smaller capacitors could be used, but to provide good filtering, the necessary series resistance (R_3) would be too big. The other extreme would be to use larger capacitors, but this would increase cost. We will select one filter capacitor with two-50 μF units, rated at 150 VDC for C_1 and C_2 since we are not spending any money.

The filter resistor R_3 and the output filter capacitor C_2 form an integrating network which in effect sums up the applied voltages and discriminates against the high frequencies. This network is actually a low pass filter and will be used to attenuate the power supply's ripple frequency. In order to attenuate sufficiently, the integrator should have a t/T ratio of 0.1 or less, a value considered to be long with regard to time constants. The large T represents the charge or discharge time of the capacitors, while the small t represents the period of the applied signal frequency, 60 hertz in our case. The following formulas will be used to calculate the specific value for resistor R_3.

(1) $T = R_3 \times C_2$	(1) value of filter resistor times the output capacitor's value
(2) $t = \dfrac{1}{f}$ or $\dfrac{1}{60}$	(2) the period, left in this form for calculation ease
(3) $t/T = 0.1$	(3) given ratio for our desired attenuation

Now, if we substitute equations (1) and (2) into equation (3)

$$0.1 = \frac{1}{60\,R_3\,(50 \times 10^{-6})}$$

and solve for R_3 we have:

$$R_3 = \frac{1}{6\,(50 \times 10^{-6})} = \frac{1 \times 10^6}{300} = 3.33 \text{ k ohms}$$

How about that for luck! The value of 3.33k ohms calculated equals the EIA standard 20% value of 3.3k ohms. We will use this value since the next larger 20% tolerance resistor is 4.7k ohms. However, before the 3.3k resistor can be bought, its power rating must be determined. We will be using the formula $P = I^2R$ for our power calculations. Note in Figure 7-4 the plate current of the 12AV6 (1.2mA) and the screen grid current of the 50C5 (8mA) both flow through R_3. The total expected current flowing through R_3 is 9.2mA and this squared equals 0.8464. If we plug this I^2 value into the formula for power we have:

$$P = I^2R_3 = 0.8464 \times 3.3k\Omega = 0.2793 \text{ watts}$$

Since ¼ watt equals 0.25 and our calculations prove to be greater than this value, a ½ watt resistor of 3.3k ohms must be procured for the filter resistor R_3.

7.1-2   Safety Design, Etc.

The power supply circuitry will now work, but we neglected safety design in our AC/DC power supply, so now to R_4 and C_3. Resistor R_4 and capacitor C_3 must do two important things; first to provide our safety feature so the circuitry and chassis grounds are not at the same potentials and second, to get rid of any noise existing between the two grounds.

We will examine C_3, the noise bypassing capacitor first because its value plays an important part in determining the resistive value for R_4.

The floating ground (the circuitry common isolated from chassis common) component selection is somewhat governed by the Underwriters Laboratory because they require that C_3 will *not* pass 5mA of current at the rated 117 VAC line voltage. Along with this UL information we will use two formulas: $XC_3 = \dfrac{ET}{IT}$ solving for the minimum capacitive reactance of the noise bypass capacitor and $XC_3 = \dfrac{1}{2\pi fC}$ solving for the value of C_3. Application is as follows:

$$XC = \frac{ET}{IT} = \frac{117}{0.005} = 23.4K \text{ ohms}$$

substituting XC of 23.4K ohms into the second formulas we have:

$$C_3 = \frac{1}{2\pi fXC} = \frac{1}{6.28 \times 60 \, (23.4 \times 10^3)} = 0.113 \ \mu F$$

The maximum value of C_3 was determined in our calculations since we used the minimum capacitive reactance value. We will select the closest in value mass-produced capacitor for C_3, but unlike the resistors we previously ordered we will select the next smaller standard value. This value is 0.1 μF. Since capacitor C_3 is not an electrolytic capacitor as indicated on the schematic diagrams, we need not worry about its polarity, only its working voltage. The working voltage for the noise bypass capacitor should be sufficient enough to safely handle the 165 volts peak or maximum AC line voltage value. A 200-volt rating should be just about right for all concerned. Now the manufacturer can buy 0.1 μF, 200 volt capacitors.

Earlier I mentioned that C_3 would play an important part in determining the value for the isolating resistor R_4. The reason for this is because R_4 is made ten times the capacitive reactance value of C_3 in order to obtain a good noise by passing between the two grounds. If we follow specific formulas we can determine the resistive and power values for R_4 as shown.

$$R_4 = 10 \, (XC_3) = 10 \, (23.4 \times 10^3) = 234K \text{ ohms}$$

The nearest EIA 20% value is 220K ohms, the value we will use to determine its wattage, (the next highest 20% value was 330k ohms).

$$P = \frac{(E \text{ max})^2}{R_4} = \frac{(165)^2}{220 \times 10^3} = 0.124 \text{ watts}$$

R_4 can be ordered. It is a 220K ohm, ¼ watt resistor. This completes the safety and noise design considerations for component selection necessary to make an AC/DC power supply.

7.2   VOLTAGE AMPLIFIER STAGE

There are two principal purposes for electronic circuits in audio amplification. The first is a power amplifier whose circuitry allows it to deliver appreciable audio power to loud speakers. This particular stage will be developed later in full detail. The second is a voltage amplifier whose purpose is to build-up applied signal currents to a level suitable for driving the final or power stage.

The voltage amplifier stage involves working with a family of curves for that particular tube to determine obtainable gain, maximum output and distortion, for any load and plate voltage. Operating points, grid voltage, or biasing, load lines, and mathematical calculations and all of the other necessary considerations will be purposely neglected, since if covered here, they would be duplicated in designing the power amplifier stage. We shall, therefore, design the voltage amplifier stage based upon educated guesses and general rules of thumb. The components required to breadboard this circuit are shown schematically in Figure 7-5.

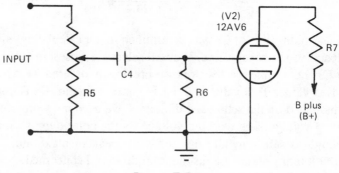

Figure 7-5

7.2-1   Volume Control

The adjustable resistor R-5, illustrated in Figure 7-5, serves as the volume control or gain adjust for our amplifier. This potentiometer allows the operator to adjust the signal input level from zero to maximum or anywhere in between. Circuitry having low bias and high gain is best suited for a volume control, preferably in the earliest possible audio amplifier stage. This control is positioned just right in our circuitry and to avoid any cost decision-making for the manufacturer at this point, we will use a 500k ohm, molded carbon composition, two watt or under.

7.2-2   Grid Load Resistor

The grid load resistor, R_6, must be considered at this time because its value will be used to select the value for capacitor C_4. For cost reduction and simplicity of the voltage amplifier, a biasing method which will prevent loading of the preceding input stage and will limit grid current through the tube will be used. Contact potential bias is the answer. A common value for this resistor is 10 meg ohms as specified in your tube manual, but its wattage is not given, only the maximum possible voltage of 1.5 volts. Now, don't get discouraged yet, this is a valuable bit of information because:

$$P = \frac{E^2}{R} = \frac{(1.5)^2}{10 \text{ meg } \Omega} = \frac{2.25}{1 \times 10^7} = 2.25 \times 10^{-7} \text{ watts}$$

This wattage value is actually 0.000,000,225.

As you can see, a ¼ watt resistor would be fine.

7.2-3   Coupling Capacitor

Now that the grid load resistor of 10 meg ohms, ¼ watt, has been determined, we can concentrate on the value for the coupling capacitor C_4. We know that C_4 will block any DC voltage possible from the signal input source, thereby protecting the grid of V_2, and we know it will pass the desired alternating signal to the grid of V_2 to cause amplification. One important piece of data we don't know is what the lowest frequency we or the manufacturer would like to amplify. This must be identified so that we can make certain in our calculations that this frequency will not be attenuated. The maker has the choice from 20 hertz to who knows. Well, for the sake of design progress, we will say the lowest frequency to be amplified will be 100 hertz. As a general rule of thumb, the capacitive reactance of C_4 must be one-tenth the value of the grid load resistor value previously given. This will insure the low frequency amplification.

Since R_6 equals 10 meg ohms then:

$$XC_4 = \frac{1}{10} \times 10 \text{ meg } \Omega = 1 \text{ meg ohms}$$

Now, in order to find out the value of capacitance for C_4 we simply plug in the 1 meg ohm resistive value into $Xc = \dfrac{1}{2 \pi fC}$ solving for C.

$$C_4 = \frac{0.159}{Fxc_4} = \frac{0.159}{(100)(1 \times 10^6)} = 0.00159 \times 10^{-6}$$

The nearest standard value is 0.0015 mfd. We will order this capacitor with a 200 Vdc rating, mainly for safety reasons, although it would depend upon the dc potential expected from the input circuitry.

7.2-4   Plate Load Resistor

The plate load resistor (R_7) can generally be determined by multiplying the rp of V_2 by 4 or 5 times. This method still allows us to obtain good gain from the 12AV6 without plotting its specifics on a family of curves. The rp listed in the tube manual for V2 is 80K ohms, therefore the plate load resistor is:

$$R_6 = 5 \times 80 \text{ K } \Omega = 400 \text{ k ohms}$$

The closest EIA, 20% value is 470k ohms and for a 10% resistor the value is 400k ohms. We will pay the additional pennies for the 10% resistor and get a 400k ohms resistor, but at what wattage? The formula $P = \dfrac{E^2}{R}$ will again pull us out of a dilemma, providing we know that the B + voltage will be about 80 volts DC. This is roughly for calculation purposes, the plate voltage supply value with current flowing, but no signal applied.

$$P = \frac{E^2}{R} = \frac{(80)^2}{400k\ \Omega} = \frac{6400}{400k\ \Omega} = 16 \times 10^{-3} \text{ watts}$$

This wattage is less than ¼ watt, but for safety reasons R_7 must be a 400k ohm, ¼ watt resistor and thus completes the voltage amplifier design.

7.3   POWER OUTPUT

The last stage will be designed in much the same manner as the previous two, except that now we shall share with you calculations based upon plate characteristic curve data. Not many general risks or educated guesses will be employed. Figure 7-6 illustrates the circuitry components we will design for, but first see if you can provide the names of purposes for components C_5, R_8, and the primary of T_1 You may not be able to name R_9 since this particular circuitry-component arrangement was not used in the power supply or voltage amplifier stage.

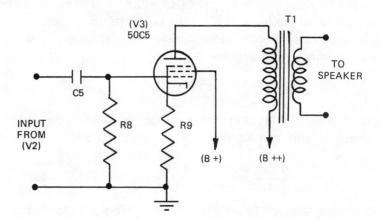

Figure 7-6

You should have called C_5 a coupling capacitor, R_8 a grid load resistor, R_9 the cathode bias resistor, and the primary of transformer T_1 the plate load for V_3.

The transformer is an impedance matching device which merely reflects back across to the primary the load impedance placed across its secondary terminals. This is in turn multiplied by the impedance or turns ratio of the transformer. Most audio output transformers of the type we are interested in are selected solely upon one value of speaker impedance. Transformer manufacturers make additional output speaker-impedance matching values. Normally these values are 2, 3, 4 or 8 ohms. What we must determine now is what specific speaker impedances we want.

In order to help the individual eager to build this amplifier and to provide the required data, we need a Stancor output transformer #A3332. The particular data for this transformer is given in Chart 7-4 and will be used to begin the power amplifier's design.

Primary Impedance	Secondary Impedance	Wattage
2000 ohms	3.2 ohms	3

Chart 7-4

The first step is to determine the turns ratio of the transformer suggested by Stancor to be used with a 50C5. The formula we will use states that the primary impedance equals secondary impedance times the turns ratio squared, therefore:

$$\text{IF} : zp = Z_s N^2$$

$$\text{Then } N = \sqrt{\frac{zp}{zs}} = \sqrt{\frac{2000}{3.2}} = \sqrt{625} = 25$$

The reflected load (RL) is determined by squaring the turns ratio and multiplying it by the second impedance matching value. Hence, the value reflected

$$RL = N^2 Z_s = (25)^2 (3.2) = 2000 \text{ ohms}$$

Note the closeness in calculated reflected impedance and that of the selected transformer. Yes, these values are not coincidental, but the same theory of design holds true.

The reflected impedance value will be used to establish the AC load line for the output tube. This value is actually the AC resistance and not the DC resistance of transformer. T_1. The DC resistance will be little help to our design since its resistance is quite low. In such cases we draw the DC load line vertically from the point representing the expected plate voltage potential. In our case, the DC load line reference is about 120 volts as shown on the family of curves graph in Figure 7-7.

The one line that perhaps is most important to us is the AC load line. From this line, values for determining stage gain, power output, biasing limitations, and tube-circuitry design specifics will be determined. One reference point is the voltage applied to the plate with absolutely no current flowing. For design purposes we will assume 200 volts is the no load power supply voltage applied to the plate of V_3. The other reference point through which the AC load line will be drawn is the value of current obtained when dividing Eb by the primary impedance of T_1.

$$I_b = \frac{Eb}{RL} = \frac{200}{2000} = 100 \text{ mA}$$

The AC load line as illustrated in Figure 7-7 has a plate current of 100 mA and a corresponding plate of 200 volts.

The tube manual gives V_3's maximum wattage, also called plate dissipation, of

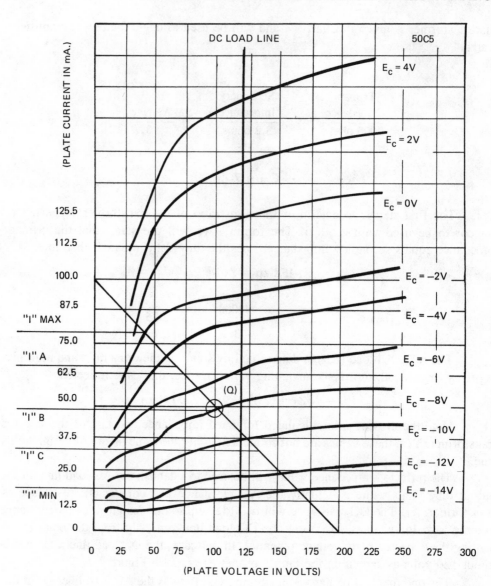

Figure 7-7

7 watts. We must make certain that our design calculations do not exceed the power limitations given. The zero signal or no signal plate current value of 49mA as suggested by the tube manual will be used to determine the plate dissipation for our amplifier stage. If a line is drawn horizontally from the 49mA no signal I_b current value to the AC load line and from that intersection straight down, we can use that information to calculate or approximate the stage wattage. As you can see from Figure 7-7, the plate current of 49mA provides a plate voltage of 105 volts.

Using this data in the formula $P = IE$ we have:

$$P = IE = 40mA \times 105V = 5.145 \text{ watts}$$

This is a good sign since we did not exceed 7 watts. In order to proceed, we need to select the grid biasing voltage that will represent the potential necessary to provide the no signal plate current. The biasing potential selected is the Ec value closest to the no signal current value. Needless to say, the value selected is -8 volts. If the plate dissipation rating was exceeded, the grid bias voltage would have to be increased and in that case the new bias voltage would be -9 or -10 volts, depending upon power calculations.

The Ec $= -8$ volt point of intersection made with the load line is called bias voltage, quiescent, or Q point. In other words, it is the operating point for the output amplifier.

The voltage gain (K) for the 50C5 is not given in the tube manual, but we can calculate its gain by using graphical data obtained when reusing Figure 7-7. Using the Q point as our reference, 1, 2, or 3 grid voltage values on both sides of the EC $= -8$ volts will provide us with a change in grid voltage (ΔEc), with a corresponding change in plate voltage (Δ Eb). For our purpose a 4-volt peak or two-grid voltage value in both directions will be used. Since we have an 8-volt peak to peak change in grid voltage we can draw the straight lines down from the points where they cross the AC load line. As illustrated in Figure 7-7, the grid voltages from -4 to -12 volts provide a voltage range from 60 to 150 volts for the plate voltage. Now that we have the change in both values we can determine the gain.

$$K = \frac{\Delta Eb}{\Delta Ec} = \frac{150-60}{12-4} = \frac{90}{8} = 11.25$$

The amplifiers wattage, if you recall, was to have been 1 watt and now is the time to determine what grid voltages for V_3 will provide the desired wattage rating. We will begin using ohms law solving for the output or secondary voltage. The value for the secondary impedance will be changed from 3.2 ohms to 4 ohms to aid in calculations.

$$Es = \sqrt{PZs} = \sqrt{1 \times 4} = 2 \text{ volts rms}$$

We will now determine the voltage for the transformer primary. Since the turns ratio is 25, the primary voltage should be 2 times 25 or 50 volts rms.

$$Ep = NEs = 25 \times 2 = 50 \text{ volts rms}$$

If we convert the primary voltage rms value to peak, the value will be:

$$Ep = Erms \times 1.414 = 50 \times 1.414 = 70.7 \text{ volts}$$

The grid voltage necessary to provide 1 watt of power is found when the peak plate voltage of V_3 is divided by the gain of V_3.

$$Ec \text{ (peak)} = \frac{Eb \text{ (peak)}}{K} = \frac{70.7}{11.25} = 6.2 \text{ volts peak}$$

The grid circuit voltage must vary from about -2 to -14 volts or 12 volts peak to peak to provide the required 1 watt output. No signal clipping will occur with this range of grid voltages, but if you examine the load line you will see a larger signal swing would go past the knee of the -2 volt grid curve.

7.3-1   Output Harmonic Distortion

Distortion in an amplifier occurs when the output waveform differs in some respect from the input waveform. This is an affect of nonlinearity. The general groupings of distortion generally worried about are: (1) amplitude, (2) intermodulation, (3) frequency, (4) phase, (5) transient, (6) voltage, (7) frequency modulation, and (8) harmonic distortion. The type of distortion we will examine is that of the harmonics.

Harmonics themselves are not necessarily displeasing since all musical instruments produce harmonic sounds. Harmonic frequencies are produced when sine wave signals, audio signals in our case, have multiples of the fundamental frequency. For example, output signal harmonics in an amplifier if a 100-hertz signal were applied to the input may consist of the fundamental frequency at 100 hertz (first harmonic), a second harmonic at 200 hertz, a third harmonic at 300 hertz and a fourth harmonic at 400 hertz, etc. This progression would continue, but in our attempts to show how to figure stage distortion, we will limit ourselves to the 4th or 5th harmonic. In fact, we will use something called the 5-point system to determine the percent of harmonic distortion present in our amplifier.

Before we begin, take time to get a paper clip and mark the page where Figure 7-7 is illustrated. The family of curves for the 50C5 and its AC load line is the source for our mathematical calculations.

The 5-point system is so called because we first assume the input signal applied to the control grid of V_3 will be a perfect sine wave, then we divide the plate current values from maximum (about 80mA) to minimum (about 15mA) into 5 equally spaced points. Figure 7-7 refers to these points as I max, IA, IB, IC and I min. When following these current values to the AC load line and from that point straight down, reference points for their particular voltages can be found. If we have done everything well, each of the 5 divisions of plate current values should line-up with the E_c grid voltage illustrated. Check your readings with those shown in Chart 7-5 which has been provided for quick reference data input in our calculations.

Ec voltage	Ib current
−2V (Ecmax)	80 mA (I max)
−5V	63 mA (IA)
−8V (Q point)	47.0 mA (IB)
−11V	36 mA (IC)
−14V (Ec min)	15 mA (I min)

Chart 7-5

The formula used to determine the percent of second harmonic distortion is:

$$D_2\% = 75 \frac{(I\ max + I\ min - 2\ IB)}{(I\ max - I\ min + IA - IC)}$$

OR

$$D_2\% = 75 \frac{[(80 + 15 - 2)\ (47.0)]}{(80 - 15 + 63 - 36)} = 0.81$$

The second harmonic distortion is 0.81 percent.

The third harmonic formula is:

$$D_3\% = 50 \frac{[(I\,max - I\,min\,2\,(IA - IC)]}{(I\,max - I\,min + IA - IC)}$$

OR

$$D_3\% = 50\ \frac{[(80-15-2)\,(63-36)]}{(80-15 + 63-36)} = 5.95$$

The third harmonic distortion is 5.95 percent.

The fourth harmonic formula is:

$$D_4\% = 25 \frac{[(I\,max + I\,min - 4\,(IA + IC) + 6IB)]}{(I\,max - I\,min + IA - IC)}$$

OR

$$D_4\% = 25 \frac{[80 + 15-4\,(63+36) + 6\,(47)]}{(80-15 + 63-36)} = -5.15$$

The fourth harmonic distortion is negative 5.15%.

These calculations could be continued, but they really are a waste of time for us since the higher order harmonics will be considered unimportant, after the 4th harmonic value has been determined. If this assumption is made, the total harmonic distortion for the amplifier can be determined using this formula:

$$DT\% = \sqrt{D_2{}^2 + D_3{}^2 + D_4{}^2}$$

OR

$$DT\% = \sqrt{(.81)^2 + (5.95)^2 + (-5.15)^2}$$

$$DT\% = \sqrt{0.6561 + 35.4025 - 26.5225}$$

$$DT\% = \sqrt{9.5361} \approx 3.08869$$

The total harmonic distortion is about 3 percent, a value of respectable quality. We will now continue with our design of component values for the cathode bias resistor (R9), the grid load resistor (R8), and the coupling capacitor (C5).

7.3-2   Cathode Bias Resistor

Since the no signal grid voltage of -8 volts was used in design, the Q point will show us the operating point plate voltage and plate current. Again, referring to Figure 7-7, the Eb value is 105 volts and the Ib is 49mA. It is at these particulars the cathode resistor must develop the bias of -8 volts. However, we must note that two currents flow through R9. These two currents are the screen current of 8mA and the plate current of 50mA (Chart 7-3). Therefore, using ohms law we have:

$$R_9 = \frac{Ec}{Ib + Isg} = \frac{8}{58mA} = 139.6\ ohms$$

The nearest EIA 20% value is 150 ohms, but you could obtain a 5%, 130 ohms resistor also. Its wattage is determined by:

$$P = (Ib \times Isg)^2 + R_9 = (.058)^2\ 150 = .5046$$

The designed wattage rating would be ½ watt, but we want to insure that the resistor can safely handle the combined plate and screen currents, so a 150 ohm, one watt resistor will be used.

7.3-3   Grid Load Resistor

The cathode biasing arrangement in our amplifier circuitry makes determining the value for the grid load resistor (R_8) a snap because the tube manual suggests a 500k ohm resistor be used. Since the grid will not be driven into the positive region (only up to -2 volts) and since we want good low frequency coupling, the nearest EIA value below the suggested 500k ohm value is selected. Based upon these points of design knowledge, we will use a 470-ohm, ½-watt resistor for R_8.

7.3-4   Coupling Capacitor

C_5 is selected based upon the fact that the desired frequency response begins at a minimum of 100 hertz. To avoid any attenuation at this frequency the reactance of C_5 must be one-tenth the grid load resistor's value. Since R_8 equals 470k ohms, the capacitive reactance should be 47k ohms at a frequency of 100 hertz. By plugging this into the following formula we have:

$$C_5 = \frac{1}{2\pi f X_C} = \frac{.159}{100\,(47 \times 10^3)} = 0.0338 \text{ mFd.}$$

The nearest standard value is 0.033 μF. Its voltage rating will be 200 volts since the plate supply voltage is about equal to the rms AC line voltage.

This completes the design of the 3-tube amplifier which just happens to be part of the required circuitry for a common AC/DC superheterodyne radio.

7.4   SIGNAL ANALYSIS

The individual values assigned to the nine resistors, five capacitors, one transformer, and three vacuum tubes illustrated in Figure 7-8 help us to recognize them physically in the AC/DC amplifier. Their purpose, often related to their name, can also help identify the components after we know their function. The hows of circuitry operation, however, are of equal importance to those of us who enjoy troubleshooting using logical, educated methods. It will be the intent of this section to illustrate how these components, specifically arranged in the AC/DC amplification circuitry, actually accomplish its design purpose.

The power supply circuitry will not be analyzed, only the voltage amplifier and power output circuitry. We must first assume that the power supply does work and that the B + and B + + voltages, without *any* current flowing, will be 200 volts. It is from this 200-volt potential that our references, when current is flowing, will be made. Throughout this portion of our signal analysis, electron current or movement will be indicated by arrows going from negative to positive.

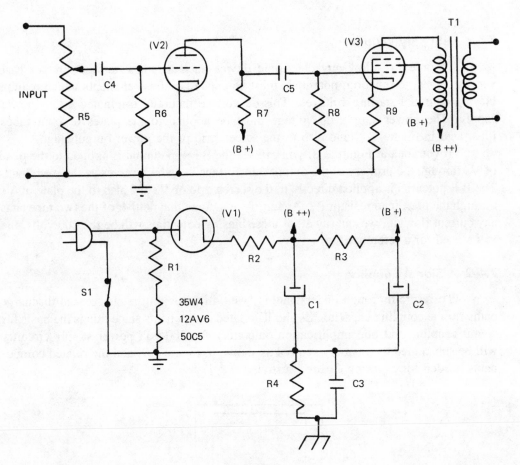

Figure 7-8

COMPLETE PARTS LIST

Resistors

R1	150 ohms, 5 watt
R2	*56 ohms, ½ watt
R3	3.3k ohms, ½ watt
R4	220k ohms, ¼ watt
R5	500k ohms, 2 watt
R6	10 meg ohms, ¼ watt
R7	400k ohms, ¼ watt
R8	470k ohms, ½ watt
R9	150 ohms, ½ watt

*68 ohms, ½ watt in 5 or 10% tolerance

Capacitors

C1 50 μF @ 150-200 volts
C2 50 μF @ 150-200 volts
C3 0.1 μF @ 200 volts
C4 0.0015 μF @ 200 volts
C5 0.033 μF @ 200 volts

Transformer

T1 Stancor #A3332

Vacuum Tubes

V1 35W4
V2 12AV6
V3 50C5

7.4-1   Supply Voltages

In order to allow any electronic device such as transistors, or in our case vacuum tubes, to function, potentials must be applied to those elements which require biasing or initial starting voltages. This dc voltage naturally originates in our power supply. The power supply dc voltage must be applied to the plates of both tubes illustrated in Figure 7-8, and also to the screen grid of the power output stage.

If you look at Figure 7-8, you can see the B++ potential is applied to the plate of V_3 through the primary of the output transformer T_1, which serves as the plate load. The B+ potential is applied directly to the screen grid of V_3 and also to the plate of V_2 through the plate load resistor R_7. Again, if we assume that neither of the two tubes has any current flowing, we can say all of these biasing potentials will be the 200 volts, the value used for a reference.

7.4-2   Signal Coupling

To insure thorough understanding, we will focus on the charge and discharge paths for the coupling capacitor C_5 and its related components since this is the basis for signal coupling and our amplification outcome. The 200-volt power supply circuitry will be substituted by a battery symbol as shown in Figure 7-9, and the related components needed for charging C_5 are illustrated.

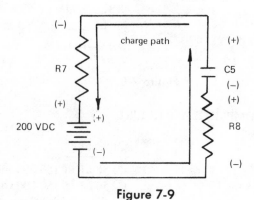

Figure 7-9

At the first instant of time, C_5 charges as indicated by the electron current path. During this period of time, the voltages drop across the plate load resistor R_7 and the grid load resistor R_8 will be maximum because at this time the electron movement will be maximum. As C_5 receives its full charge, the voltages dropped across resistors R_7 and R_8 will decrease since the current flowing through them is decreasing. At the end of its charge time (theoretically 5 time constants), capacitor C_5 is fully charged to, or almost to, the 200-volt source potential. This will remain charged to the 200-volt potential until a discharge path, illustrated in Figure 7-10, is provided.

The discharge path for C_5 is provided through V_2 during the time it is allowed to conduct. V_2's internal resistance decreases when it conducts, thereby allowing C_5 to

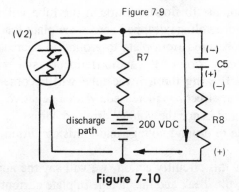

Figure 7-9

Figure 7-10

discharge as indicated by the electron arrow path. The discharge time depends upon the value of R₈ and the internal resistance of the vacuum tube when it is allowed to conduct. Figure 7-9 illustrates what happens when a negative signal is applied to the control grid of V₂, while Figure 7-10 shows what occurs when a positive signal is applied to the control grid.

7.4-3   Voltage Amplifier

Before any audio frequencies are applied to the voltage amplifier's control grid, we will examine what takes place when V₂ has no signal applied. Since the tube has B+ on its plate and its cathode is more negative than the plate, plate current (Ib) will flow through the tube and through the plate load resistor R₇. The no signal applied tube current flow is illustrated in Figure 7-11.

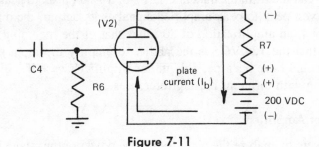

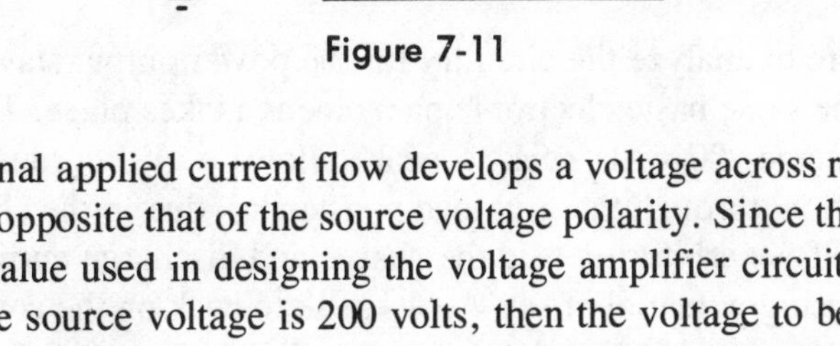

Figure 7-11

This no signal applied current flow develops a voltage across resistor R₇ whose voltage polarity is opposite that of the source voltage polarity. Since the normal operating plate voltage value used in designing the voltage amplifier circuitry was about 80 volts, and since the source voltage is 200 volts, then the voltage to be dropped across R₇ would equal the difference of 120 volts dc. If we divided this voltage difference by the resistance of R₇, we could use this current value for another reference; hence, the approximate tube current with no signal applied would be 0.3mA.

The tube plate current value will increase or decrease, depending upon whether the applied audio signal is positive or negative. The positive portion of the audio frequency wave will reduce the vacuum tube's internal resistance, allowing the tube to conduct, and in doing so, the Ib flowing through the tube will increase. When this happens, there will be more voltage dropped across the plate load resistor R_7 and will cause the plate voltage measured from plate to ground to decrease.

When the negative portion of the audio frequency sine wave is applied to the control grid, the current flowing through the tube will become smaller because the tube's internal resistance in effect has increased. With a negative applied to the control grid, less current flows, and hence the voltage dropped across R_7 will decrease. If you were to measure the plate to ground voltage under this condition, you would find that Eb increased.

In order to clarify this circuitry action, we will say the applied signal will be 1 volt peak to peak or 0.5 volts peak and the change in plate current caused by the 1-volt peak to peak signal applied will, for our purposes, cause a 0.1mA peak or 0.2mA peak to peak current change in the tube. Chart 7-6 illustrates these changes of plate current and plate voltage when one complete audio frequency sine wave is applied to the control grid of the voltage amplifier stage.

E_c	I_b	$(E_s - E_{R7})$	$= E_b$
0.5 vp	0.4mA	200−160V	40
0v	0.3mA	200−120V	80
0.5 vp	0.2mA	200−80V	120

Chart 7-6

As you can see from the data in Chart 7-6, a 1 volt peak to peak input signal will provide an 80 volt peak to peak amplified signal as its output, the difference between 120 and 40 volts, an amplification of 80 times that of the incoming signal. The tube manual states that the 12AV6 has the ability to amplify the input signal 100 times. Greater gain would have been possible in our amplifier's case if a higher B+ voltage were used along with different biasing references.

7.4-4   Power Amplifier

If we were to analyze the circuitry for the power output stage involving V_3, we would see that the same basic electronic phenomenon takes place. The signal amplified in the previous stage, 80 volts peak to peak, allows a grid current of about 0.17mA peak to peak to flow through the grid load resistor R_8, due to the charge and discharge of capacitor C_5. If we arbitrarily said the charge and discharge current of C_5 caused an instantaneous biasing potential of say 4 volts peak to peak on the control grid of V_3, we could examine the vacuum tube data shown earlier in Figure 7-7 to see what would happen. As you can see, the values for Ec would vary from −6 volts for the positive alternation of the input signal to −10 volts for the negative signal alternation. The change in plate current would be from about 35mA to 62.5mA, which results in a plate voltage change of 130 to 85 volts or 45 volts peak to peak.

If you analyzed the voltage amplification for this tube it would be determined that a 2-volt peak input signal provided about 22.5-volts peak output signal, a total voltage gain of about 10. Since V_3 is not a voltage amplifier we should not become alarmed because its purpose is for current gain. For comparison, the input current of 0.17mA peak to peak (C_5 charge and discharge) causes a plate current change from 35 to 62.5mA or 27.5mA peak to peak. This represents a total current gain for the power output stage of over 160.

Perhaps this circuitry analysis will help shed some light on the charge-discharge, current increase-voltage decrease, and current decrease-voltage increase signal amplification process common in all electronic circuitry. Figure 7-12 illustrates the signal path for the entire amplifier. Also shown is the signal phase shift resulting from the increase in plate current with a positive input alternation causing a decrease in plate voltage across the vacuum tube. This is why the control grid voltage signal is 180 degrees out of phase with the output plate signal.

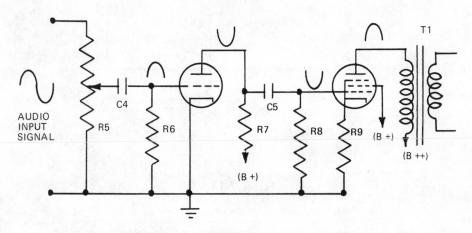

Figure 7-12

7.5   RESISTANCE ANALYSIS

Troubleshooting electronic devices with an ohm meter often proves to be quite helpful, providing we know in advance what the resistance reading should be. Application of this section to your everyday troubleshooting methods will enable you to recognize what resistance readings are acceptable. Again, the now quite familiar AC/DC amplifier circuitry will be examined; in particular, we will make our own specific resistance measurement chart.

Figure 7-13 will be used to aid in our analysis. Those pin numbers corresponding to the elements within the specific vacuum tube are illustrated as are the resistance values for the resistors in the circuitry. Resistor R_5 will be neglected since it connects directly to the input source.

Since there are two ground common points in the AC/DC amplifier circuitry we

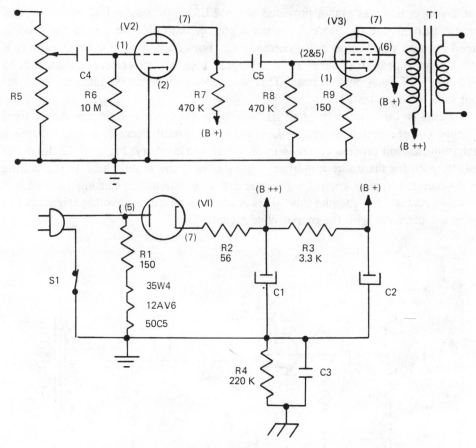

Figure 7-13

will have two different resistance values measured depending upon which common point was used. For example, measuring from the circuitry ground instead of chassis ground, the anticipated resistances for both control grids and both cathodes of V_2 and V_3 would be as indicated in Chart 7-7.

V_2		V_3	
pin (1)	10 MΩ	pin (2)	470k Ω
pin (2)	0Ω	pin (5)	470k Ω
		pin (1)	150 Ω

Chart 7-7

The only resistor between pin 1 of V_2 and ground is 10 meg ohms; therefore, it is the anticipated resistance to be measured. The remaining pin numbers shown in Chart 7-7 have the resistance of that resistor which is connected between the pin number and the common reference point. If, however, the chassis ground point were used, the resistive value of R_4 (220k Ω) would have to be added to those resistance values given in Chart 7-7.

The plate to common reference resistance values for both tubes was purposely avoided because their specific measurements must be taken between their plate pin numbers and the point at which DC originates. In order to determine these resistances, pin 7 of V_1 is used for the one lead of the ohm meter that was previously attached to one of the grounds. For example, the anticipated resistance from pin 7 of V_2 to pin 7 of V_1 would be the sum of the series resistors $R_7 + R_3 + R_2$. Chart 7-8 shows the resistance readings for the remaining specific tube pin numbers.

V_2		V_3	
pin (7)	473,356 Ω	pin (7)	56 Ω + T_1 primary resistance
		pin (6)	3,356 Ω

Chart 7-8

These same basic concepts hold true in all electronic circuits and will, when applied, help you master Electronics. If someone were now to ask you what Electronics is all about, take a deep breath and say, "three circuits, five components, and a whole bunch of facts based upon component characteristics, circuitry design, type of source, frequency, or voltage and many rules, formulas and laws."

INDEX

(All references are to section numbers)